Title: E-Mental Health: A Guidebook for Psychiatrists
Authors: Melvyn WB Zhang, Roger CM Ho

Contents:

Chapter 1: Overview of Internet and Smartphone Technologies

Introduction to Smartphone Technology for Healthcare Professionals (for patient's care)

Over the last few years, numerous studies have been published demonstrating the increased adoption of smartphone and smartphone-related applications by healthcare professionals (Garritty & Eman, 2006). Smartphones represent a new modality of technology that offers more than the conventional modalities of mobile technology. Smartphones are equipped with immense computing capabilities that allow individuals to access the Internet at their convenience. A Smartphone application is a computer program that is specially designed to run on smartphones and tablet computers, usually serving a specific purpose. In fact, the healthcare system itself is inherently mobile in nature, encompassing consultations and treatment at a wide variety of locations, including clinics, inpatient wards, outpatient specialist services, emergency departments, operating theatres, intensive care units and even laboratories. To complete their day-to-day tasks, healthcare professionals must be mobile while simultaneously communicating effectively and collaborating with colleagues across disciplines (Abu, 2012). Hence, it is not surprising that the smartphone adoption rates have increased among healthcare professionals, who realise the invaluable additions these tools can provide to their daily practices.

Previous systematic reviews have examined 23 surveys on the use of PDAs by healthcare professionals across several countries. The key findings include that younger clinicians are more likely to use this technology and that PDAs were most commonly utilized by family doctors and general practitioners. In particular, large practice- and hospital-based clinicians are more likely to use mobile technology (Abu, 2012). Thus, conventional technological modalities are currently being replaced by smartphones, which combine the basic functions of a pager, a cell phone and a PDA with even more sophisticated capabilities. This major revolution in the medical industry may have begun in 2007 with the release of Apple's iPhone. Since then, the smartphone market has advanced rapidly (Karl, 2012). Another pivotal change was the launch of the Apple Application Store in July 2008 (Karl, 2012), which enabled users to download smartphone-based applications to add additional capabilities to their smartphones. A recent survey performed by Manhattan Research in 2009 (Abu, 2012) showed that approximately 64% of the clinicians in the United States have used smartphones, compared with 30% in 2001. Other research statistics reflect the tremendous increase in the number of smartphone applications being downloaded, from 300 million in 2009 to over 5 billion in 2010 (Mobile Future, 2010).

With these technological advances, clinicians are no longer confined to using individual workstations or computers on wheels (Abu, 2012) when accessing hospital information systems, electronic health and medical records, laboratory results, and/or the latest evidence-based information to help them with clinical management. With the increased adoption of smartphones, clinicians have easier and better access to patient information, which should improve clinical care (Abu, 2012). Additionally, new software applications are increasingly tailored to the needs of clinicians, particularly resources for evidence-based information. In addition to clinicians, patients can benefit from this technology to monitor their own conditions. Previous studies have demonstrated the efficacy of smartphone use in disease prevention and self-monitoring as well as in the management of chronic diseases (Marshall, 2008).

Previous studies (Abu, 2012) have examined differences in the utilisation of healthcare applications among healthcare professionals. Despite the variety of applications used, those for the diagnosis of diseases and medical calculations are the most commonly used by practicing health professionals and in studies in medicine and nursing. Previous studies (Abu, 2012) have highlighted the ways in which these new technological modalities have enhanced patient care, which will be explored in the next section.

Current Applications of Smartphone Technology (Applications) for Patient Care in Medicine

These new technological modalities have enhanced patient care in many ways. A review of previous studies highlights how new smartphone technology has enhanced patient care.

Chronic Disease Management	- Abu (2012), in a paper titled "A Systematic Review of Healthcare Applications for Smartphones," identified 6 applications for the management of chronic conditions (diabetes, cardiac rehabilitation and COPD rehabilitation) - Ozdalga (2012) identified an application called Diabeo for chronic diabetes self-management (Charpentier, 2011). This diabetes application collates information regarding self-measured blood sugar, carbohydrate counts and planned levels of physical activity. Researchers in France studied the application extensively over a 6-month period with 180 patients and noted that those patients who used the application tended to have lower glycated haemoglobin levels than those who did not use the application.
Patient Monitoring	- Ozdalga (2012) reviewed the android application iWander (Sposaro, 2010), which helps to monitor geriatric patients with Alzheimer's dementia through the smartphone's GPS, which can indicate when the patient has wandered away from his or her usual environment.
Rehabilitation	- Smartphone applications have also been useful in rehabilitation (Wu, 2011). - Patients who have difficulty attending a rehabilitation programme can now be monitored via Bluetooth with a single lead electrocardiograph while they exercise at home. The tracked information will help to create customised exercise regimens for patients who have recently suffered a recent coronary artery event or undergone angioplasty (Worringham, 2011). - In addition to the usefulness of smartphones for cardiac rehabilitation programmes, a crucial role in post-stroke

	rehabilitation has been identified. The activity levels of post-stroke patients can be monitored using shoes fitted with special sensors that direct all collected information back to a smartphone (Edgar, 2010).
Medical Diagnosis	- Bsoul (2011) demonstrated the utility of the smartphone in assisting clinicians with medical diagnoses. The diagnosis of sleep apnoea is now possible remotely using a smartphone coupled to a single-lead EEG, significantly reducing the costs involved in diagnosing sleep apnoea. - In 2011, MobiSante became the first company in the world to design and build an ultrasound diagnostic probe that was approved by the United States Food and Drug Administration (FDA). - Other current clinical applications of smartphones include their use as an ECG recording device (Oresko, 2010) or as a Doppler device to measure blood flow (Huang, 2012).
Outreach Efforts in Developing Countries	- Healthcare workers in rural Thailand are now able to use smartphones to help them treat malaria (Meankaew, 2010). Using smartphone applications, have allowed better follow-up and medication adherence rates in patients. - Healthcare workers in Kenya have also benefitted from this new, innovative technology to collate data obtained during home visits (Rajput, 2012).
Applications for the Most Prevalent Global Conditions	- A recent report from Perez (2013) evaluated mobile health applications for the most prevalent conditions based on the World Health Organisation's Global Burden of Disease. - The World Health Organisation has highlighted the following conditions as the most prevalent worldwide: iron deficiency anaemia, hearing loss, migraine, poor vision, asthma, diabetes mellitus, osteoarthritis and depressive disorder. - The results of Perez (2013) show the discrepancies between literature reviews and commercial reviews. There are significantly fewer literature reviews of applications than commercial reviews. - Although iron deficiency anaemia has the greatest global prevalence, the current study shows that there is a paucity of research examining the relevant existing applications; in addition, significantly fewer commercial applications have been developed for this condition. - In comparison, diabetes and depression, which are lower in prevalence according to the World Health Organisation, have a greater number of commercialised applications and existing literature reviews.

Overview of advancement in Smartphone technologies for education

Over the past decade, there have been massive developments in Web-based and Internet technologies. In addition, there have been even greater advances in terms of mobile phone based technologies, with the introduction of and the ever increasing popularity of Smartphones. Smartphones are a new generation of mobile technology that has created much of a revolution in the current telecommunications market (Hamid & Kavit, 2011). Smartphones are currently equipped with immense computing capabilities that allow individuals to access the Internet always on the go. Smartphones have the capabilities to facilitate more than purely voice and text-based communications. They are generally now being regarded as handheld computers, rather than just mobile telephones (Hamid & Kavit, 2011). It was perhaps the release of Apple's iPhone in 2007 that sparkled off a major revolution in the tele-communications and information technology arena. From 2007 onwards, the smartphone market developed and advanced rapidly, with Apple releasing their new iPhone 5S version recently, and competitors releasing equivalent or even superior models (Karl P, Heather W, & Kim W, 2012). What was also regarded as pivotal towards the further advancement of smartphone technologies was perhaps the launch of the Apple Application Store in July 2008 (Karl P, Heather W, & Kim W, 2012). The application store enabled users to download smartphone-based applications that allowed for added additional capabilities to the smartphone, apart from them being used just as tools for accessing the Internet. Previous statistics have shown an increase from over 300 million applications being downloaded in 2009 to over 5 billion in 2010 recently (Mobile Future, 2010). In particular, there are currently more than 7000 healthcare related applications that have been made available for download (Free C et al., 2010). Even though it was Apple who have initiated the idea of the application store in 2008, other major technological companies have also followed suit and have launched similar application stores, such as the Google Android 'Play' Store as well as the Blackberry application stores (Karl P, Heather W, & Kim W, 2012). These initiatives have enabled users of other mobile-based platforms to have access as well to various mobile applications.

For the general public, the healthcare applications that are currently available on the application stores at the moment will be able to help provide them with crucial healthcare information, advice and instructions (Free C et al., 2010). In addition, some of these applications do have an additional interactive tool, which enables individuals to monitor record and reflect their physical and psychological wellbeing (Free C et al., 2010). For the healthcare professionals, much has changed since the introduction of smartphones as well as the associated smartphone applications. Medical doctors are now able to store entire textbooks within their smartphone devices and access them when needed. In addition, they are also able to make use of resources such as medical calculators and drug formularies at the bedside to help them with their clinical diagnosis and management. More importantly, the applications have the capacity to self-update and hence, this has helped to ensure that the information within the application is always kept current (Karl P, Heather W, & Kim W, 2012). There has been a recent systematic review of healthcare application that was conducted by Abu M, Illhoi Y & Lincoln S (2012). In their systematic review of healthcare applications, articles that discussed about applications were extracted from MEDLINE. Of significance, they have identified that disease diagnosis, drug reference, and medical calculator applications have been reported to be the most useful by healthcare professionals as well as medical or nursing students (Abu M, Illhoi Y, & Lincoln S, 2012). The recent systematic review also pointed out differences in terms of usage of healthcare related applications amongst the different medical professionals.

With the advances in smartphone technologies and the different utilization in terms of healthcare applications amongst different medical professionals in mind, it is pertinent to explore more into the perspectives that particular groups of users have. It will be of interest for us to look into the perspectives of medical students and trainees towards smartphone devices and their applications, as the topic of interest in the current review article is that of the application of smartphone technologies in education.

Overview of Medical Students and Trainees' perspectives towards Smartphone

Since 2006, journals such as Academic Psychiatry have proposed that knowledge and skills in informatics will be essential towards life-long learning and modern clinical practice (Hilty DM et al., 2006). In addition, it was proposed that existing psychiatry curriculum needed to integrate with the advancement in technologies (Hilty DM et al., 2006). A previous pilot study was done, looking much into user perceptions of technology in medicine and the pilot study has demonstrated that residents and medical students in 2006 have already felt that technology skills are integral in their existing medical training. Back in 2006, the most advanced portable device was still that of a personal digital assistant, and it is not surprising that the previous study identified that medical students and residents felt that these devices (personal digital assistants) provided them with core information they required during ward rounds and in their clinical settings. Trainees felt that those forms of technologies available back then were pertinent in helping them in their clinical care of patients as compared to other conventional modalities of searching for evidence based information (Briscoe et al., 2006). Given that there have been major advances in technologies since 2006; it is of importance to look into the perspectives of medical students and trainees with regards to newer modalities of technologies such as smartphone technologies.

Studies done more recently have looked at medical students and trainees' ownership, usage and perspectives towards smartphone usage. In 2012, a questionnaire-based survey was distributed amongst the interns in the Republic of Ireland (O'Conner et al.). It demonstrated that smartphones were widely adopted and accepted and that they were being used daily by interns in performing their regular duties. Based on the results from the questionnaire study, the British National Formulary application was the most commonly used application by interns. In terms of web-technologies, Wikipedia was labeled to be the most frequent used website (O'Conner et al.). Regional survey of doctors in the United Kingdom has been previously done by Payne et al. (2012). In their study, an online survey of medical student and foundation level junior doctors was conducted within a single United Kingdom healthcare region. Their online survey involves asking doctors whether they had have access to a smartphone and if they have had used applications on their smartphone devices to help support their education and practice activities. Data pertaining to the frequencies of usage as well as the type of application used were collated. In addition, participants' perceptions about the usefulness of particular applications were also collated. In Payne et al. (2012) study, they have surveyed a total of 257 medical students and 131 junior doctors. Their study demonstrated that there is a significantly high level of smartphone ownership in the cohort that they have studied. In addition, the majority of the participants in their cohort owned between 1 to 5 medical related applications. As compared to other platforms, iPhone users were more likely to own applications. Both the medical students and the trainee doctors had rather similar usage of applications, with most of them using applications for between 20 to 30 minutes per day. The most frequently used applications include that of disease diagnosis, management and drug reference applications.

Even more recently, in 2013, Tim et al. did another variant of a questionnaire-based study looking into the usage of smartphone and its acceptability among clinical medical students. They distributed a questionnaire to clinical medical students at the University Of Birmingham, UK. A cumulative total of 361 participants took part, and of which 59% of them owned a smartphone and approximately 37% of them actually did report using the device to help them in their learning. Of importance, the students who were surveyed were generally positive towards the concept of using smartphones as future educational aids, with at least 84% of them believing that the devices to be useful. However, it is also of importance to note that approximately 64% of them felt that smartphones might be too costly to implement in the clinical educational settings. Despite the generally positive attitudes that students have had towards smartphones, they did report that it might be result in unprofessional behaviors and excessive over-dependence on smartphone in clinical settings. Other research has also looked into how mobile technology has helped to support trainee doctors' workplace learning and patient care (Hardyman et al., 2013). Thematic analysis done based on a survey of trainee doctors have demonstrated that smartphone would be of much help in the following areas: teaching and learning from observation; transition from medical student to new doctor; trainee doctors' discussion with seniors; independent practice; patient care and also accessibility to reliable information, which would in turn support confident and efficient decision making.

Past research which have been described above are largely limited to questionnaire based surveys looking into ownership and usage rates of smartphone and their applications; as well as the general perspectives of medical students and trainees towards smartphone and applications. There has been a paucity of research looking into specific intervention, for example, implementation of a specific application and evaluating perspectives towards the intervention. This gap in terms of research evidence has since been looked into recently, by Payne et al. (2014). Payne et al. (2014) did a pilot study, to investigate the impact of implementation of a hospital specific smartphone application to a cohort of British Junior doctors. The investigators created an iPhone application that contained mainly disease management and antibiotic dosing guidelines specific to a hospital and trialed the application amongst 39 foundation year doctors across a total duration of 4 months. Their results showed that participants felt generally positive towards the availability of having such an application, with 68% of them indicating that the application helped to save them much time in clinical activities.

Hence, taking into consideration the result of the previous survey questionnaire that have been conducted, as well as the pilot study done recently, it is of no doubts that students and trainees have recognized the positives of smartphone technologies in various domains of their clinical duties. It is thus of importance to clearly delineate and look into existing literature to determine how smartphone and smartphone applications have affected education in medical and surgical specialties.

Overview of Smartphone Usage for education in other Specialties

Given the generally positive attitudes and perspectives that trainees and medical students have towards smartphone and its applications, more specialties are actively embracing smartphone as an educational tool. An overview of how several specialties have utilized smartphones as an effective means in education will be described.

Pediatrics

Hawkes et al. (2012) described in their paper that pediatricians have long recognized the usefulness of smartphones. However, the existing gap in knowledge is that there have not been any studies demonstrating the usefulness of using smartphones and its application in the teaching of core clinical skills. Hence, Hawkes et al. (2012) have conducted a study, to which the main objective was to determine if a smartphone neonatal intubation instructional application, named Neo-tube, will be effective towards enhancing trainee knowledge as well as enhance their procedural skills performances in neonatal intubation. In Hawkes et al. (2012) study, a total of 20 pediatric trainees participated. Their results highlighted that, with the usage of a smartphone application in enhancing existing educational models pertaining to neonatal intubation, there was generally an increase in the overall skills score, along with a decreased in the duration of each intubation attempt. This study is certainly a key landmark study for the pediatric specialty, as it has shown that with the usage of newer modalities of technologies, there was an overall enhancement of knowledge and procedural skills performance. The results from this study are pertinent, as it does show that procedural skills education applications do indeed enhance clinical care, and that this might be an area that other disciplines could consider as well.

Ophthalmology

Hassani et al. (2013) has previously looked into smartphones usage in ophthalmology. In November 2012, they have done a search through the Apple application stores and have identified around 342 applications of relevance to ophthalmology. They have delineated the applications into 2 separate groups: one for ophthalmologists and another for patients. Even though Hassani et al. (2013) have not explicitly mentioned how these applications are used in terms of educational purposes, they have pointed out that there are currently some applications out there in the application stores that could be used as clinical devices for ophthalmologists. It could be inferred that these applications might be helpful in terms of educating and enhancing ophthalmologists' clinical skills.

Nephrology

Bhasin et al. (2013) have highlighted in their article that Chronic Kidney diseases and its complications are usually associated with significant morbidity and mortality. They have pointed out that there have been previous studies that have demonstrated that there are currently significant gaps in knowledge that residents possess, in terms of their awareness of Chronic Kidney diseases and its related complications. Bhasin et al. (2013) have emphasized in their paper the need to improve existing Chronic Kidney disease education by utilizing newer modalities of technologies, such as online websites, blogs, online modules and even smartphone applications. They have made such a suggestion in view of the current existing restrictions in terms of work-hours, and also in view of the increasing workload amongst residents. Bhasin et al. (2013) in their review paper have identified several online resources that are believed to be of value for medical students and residents. It is hoped that with the increased recognition of these online tools, medical students and residents will utilize these tools to fulfill their existing gaps in knowledge with regards to the complications of Chronic Kidney diseases. In Bhasin et al. (2013) paper, they have also pointed out the importance of recognizing the potentials of smartphone and tablet based applications. They have pointed out several nephrology applications that are available, such as calculators for nephrology-based equations, and drug information for kidney dosing. However, Bhasin et al. (2013) have emphasized, that unlike other disciplines, in nephrology, there is still currently a lack of smartphone educational tools for clinicians and medical trainees. They have highlighted that this is potentially an area of opportunity for those who are interested in nephrology education for medical students and also trainees.

Plastic Surgery
Nada H & Sudip G (2013) have in their article gave a concise summary of the uses of smartphones and its applications for plastic surgeons, in the domains of education, telemedicine and also global health. Based on their previous review, they have pointed out that despite the fact that applications are currently being developed for every imaginable use, there is still a relative limitation in the number of plastic surgery-specific applications that are made available. However, they have concisely summarized other applications and have identified at least 16 applications that would be of educational value for the plastic surgeon. They have highlighted the application named "Mersey Burns", as it is an application that is deemed to be useful for the trainees to determine the burn area percentages and helping them in the computation of the fluids that are required to be prescribed in accordance to standardized formulas, such as the Parkland Formula. Beside identification of the usefulness of smartphone-based technologies in enhancing education, they have also identified mobile-based websites, podcasts and videos and electronic books as useful educational aids.

Orthopedics
Nawfal AH et al. (2012) have pointed out in their recent paper with along with the introduction of the European Working Time Directive, trainees, especially those in the surgery discipline are currently faced with limited training opportunities. Advancement in technologies and the growth of smartphone-based technologies are thus being seen as aids to help trainees and junior surgeons make the most out of their limited time. They have pointed out and highlighted the potentials of smartphone, not only in the domain of education, but also as an invaluable tool for clinical care. Commonly used orthopedic applications include that of AO Surgery Reference, which is an application that provides immediate access to information relating to procedures, and thus could be used as a quick reference prior to any procedure. Other commonly used educational applications include that of "Zollinger's atlas of surgical operations" and "iSpineOperations", which provide trainees and junior surgeons with 3D animations of the cervical and lumbar spine procedures. In addition, the authors have pointed out that smartphone could have also value add by being of assistance to trainees in helping them complete their work-based assessment. In addition, they have highlighted that more journals do also provide podcast and video tutorials that are now accessible via smartphone devices; hence, this will be of educational value, as it

will definitely help in expanding the knowledge base of trainees and junior surgeons. Hence, it is beyond doubts that smartphone has definitely a place as an educational tool for the orthopedic specialty.

ACGME Training Programs: An overview of the research done previously in the United States

A recent study done by Orrin L & Timothy FT (2011) have demonstrated that smartphone and their applications are currently being used widely in all the current ACGME training programs. Based on their email survey about smartphone and associated applications usage, which has include a sample of 3,306 unique responses from 1397 resident, 524 fellow and 1385 attending physicians, it was noted that more than 85% of the respondents have indicated that they have had ownership of a smartphone. Of importance, their study highlighted that the most frequently requested app types by the respondents were that of textbook/reference materials, classifications treatment algorithms and general medical knowledge. In 2011, they have already emphasized that the clinical usage of smartphone and applications will likely increase, and the authors have also highlighted the need for higher-quality applications. Interestingly, out of the 134 psychiatry respondents that they have surveyed, 84.3% of them are using smartphone and around 62.6% of them have been using applications. Chi-square testing was done, which did not show an association between the level of training and the application usage; however, there were no further analysis to determine if there were differences between the individual specialties.

Pharmacology

Faye H et al. (2013) in their recent article published has clearly demonstrated that there are currently a wide number of smartphone applications that could help support users who are prescribing medications. In particular, the article, through its rigorous search, have highlighted that the majority of the current applications not only offer the potential to improve the ease and the accuracy associated with dose calculation and intravenous dose calculation and delivery; but added that the current wealth of applications in the application stores also allows users to have greater ease of access to popular pharmacological textbooks, guides and journals. The only limitation with regards to these textbook-based educational application might be the very fact that some of these applications are currently deemed to be very expensive and not affordable by all users.

Even though there are currently regulations in both the European Union, such as the Medical Device Directive (MDD) and in the United States, such as the Food and Drug administration's proposal, these directives have yet to be official legal requirements and hence the authors have cautioned users about the quality of applications. The authors have proposed that instead of waiting for the development of statutory regulations, it might be wiser for Universities and healthcare organizations to either create their own applications; or embark on peer review of applications to determine which applications are deemed suitable for users. The authors stressed that the development of in-house applications could also address the educational shortfalls and the resultant shortfalls in prescribing competencies of medical undergraduates. In addition, the authors have also highlighted that educational smartphone applications do have a great potential in improving the prescribing performance of junior doctors, given the already existing popularity of e-learning in clinical pharmacology.

Urology

Hamid A & Kavit Amin (2011) has both reviewed smartphone applications for the Urology discipline. Based on their preview review, from the applications that they have collated from the application stores, it does seemed that urologists are at the forefront of this advancement in mobile based technology, as there are currently many practical applications that could potentially be of great use for both the junior trainee as well as the senior consultant. Apart from highlighting relevant applications that will help urologists in their daily clinical practice, they have highlighted several applications that are catered specially for educational purposes. In particular, they have introduced applications that will enable urologists to have quick access to relevant information pertaining to their residency program, as well as international fellowships and a host of beneficial training resources such as textbook recommendations, video libraries and even job vacancies. Focusing more on educational needs of residents and trainees, the authors have described how the widely used Oxford handbook series, which has been crafted into an application, have potentially educational benefits as it offers quick access to relevant information for trainees. In addition, the educational needs for postgraduate examinations such as the FRCS Urology examinations have also been met through applications such as the urology flash cards series. More importantly, the authors have highlighted that there are existing conference applications in the application stores (such as the AUA and the European Association of Urology conference applications), which will provide trainees and consultants with timely updates regarding the latest developments in their field. There are also other social media based applications for urologists out in the marketplace, and the authors have pointed out that these are useful modalities for open communications and interaction between urologists.

Chapter 2: Current limitations in applications of these technologies for Psychiatry

Existing Psychiatry Applications for Patient Care

A search of the existing literature reveals that the major advances pertaining to the utilisation of smartphone technology have occurred predominantly in the following areas. This overview covers the application of smartphone as well as smartphone applications.

Schizophrenia	
Text Messaging Technology	Granholm (2011) previously attempted to incorporate the use of mobile devices into the clinical care of patients with schizophrenia. In Granholm's (2011) previous pilot study, participants received text messages daily; if they responded, advice regarding the management of their symptoms would then be delivered via their mobile devices. In the pilot study, approximately 86% of participants responded. Although there were qualitative self-reports of a reduced sense of self-distress and improved medication adherence, no significant changes in the baseline outcomes were noted. Granholm (2011) did recognise the inherent limitations of their pilot study and later proposed more user-friendly interventions using smartphone technology, and these may provide greater benefits to patients. Spaniel (2008) used text messaging to inform psychiatrists regarding the well-being of their patients. In that study, recently discharged outpatients at a high risk of relapse as well as their family members received weekly text message requests to complete an assessment of early warning signs of relapse. The results showed that when the treating psychiatrists responded to assist patients who exhibited symptoms suggestive of a relapse, there was a marked reduction in the absolute number of days of hospitalisation and overall medical costs, compared with controls.
Use of Smartphone Technology	Palmier-Claus (2012) conducted a feasibility and validity study examining the effectiveness of self-reporting psychotic symptoms using a smartphone-based software application. Palmier-Claus (2012) highlighted the current limitations of existing semi-structured interview scales; the primary limitation was recall bias when patients are asked to evaluate their symptoms over the last 7 to 28 days. Additionally, regarding existing scales, raters require training to ensure the reliability of the scores. Palmier-Claus (2012) proposed that ambulatory self-reporting in real time would have an advantage over previous methodologies with respect to these limitations. In their study, a cumulative total of 44 participants were recruited to test a monitoring system for psychosis using innovative technology. The participants were asked to complete a total

	of 14 branching self-report items that pertained to their key psychotic symptoms using their smartphones. Participants were prompted to do so by an alarm that sounded randomly six times a day for one week. Based on the researchers' analysis, the 5 items scored by participants on their smartphones showed a strong correlation with the scores obtained with the usual rating scales. These 5 items were delusions, hallucinations, suspicion, anxiety and hopelessness. The initial results were promising and demonstrated that a smartphone device could be used to assess psychotic symptoms in individuals. The main advantage of this modality is that it enables the accurate collection of participants' symptoms across various time points and situations (Palmier-Claus, 2012), which is useful in subsequent psychiatric consultations. This technology may help to identify specific relapse triggers while assessing participant responsiveness to medical treatment (particularly in the acute phase when treatment has been initiated).
	Palmier-Claus (2013) examined the possibility of integrating a mobile-phone-based assessment for psychosis into patients' everyday lives. A total of 24 patients were recruited based on the selection criterion of having a non-affective psychosis. These individuals were required to fill out a self-reported questionnaire regarding their symptoms using either text messages or a specially designed android smartphone application. The duration of the study was 6 days. The initial results from this qualitative study demonstrated that patients considered this monitoring modality to be helpful. For clinicians, the main benefit was their increased understanding of the symptoms experienced by their patients. This self-reporting modality also improved the doctor-patient relationship because the clinician had more details regarding the symptoms.
Bipolar Disorder	
Use of Smartphone Technology	Grunerbl (2014) recognised the potential of smartphone technology and performed a pilot study on the utility and efficacy of the smartphone and behavioural management devices. That study, based on a newly developed system, examined the use of smartphones and wearable devices for recognising the depressive and manic states of their patients and to identify changes in state in their patients. The initial pilot study used 4 different sensing modalities and showed that the recognition accuracy was approximately 76%.
	Faurholt (2014) tested a daily electronic monitoring smartphone software called MONARCA on 17 patients for 3 months. Patients were required to complete both the Hamilton Depressive Rating Scale and the Young Mania

	Rating Scale every two weeks. The preliminary study showed that the self-monitored depressive symptoms correlated with the scores and that objective smartphone measures such as physical and social activity correlated clinically with the depressive symptoms.
Depression	
Use of Smartphone Technology	A pilot test of the guided Mobilyze! Application identified a significant reduction in the depression scores (as assessed by the Mini-International Neuropsychiatric Interview [MINI]) and in depression and anxiety symptoms during the post-test period (as assessed by the Patient Health Questionnaires). There has been recent demonstrated feasibility of using depression application and validated questionnaire such as the PHQ-9 for the screening of depression (BinDhim 2014).
Anxiety	
Use of Smartphone Technology	Three previous randomised controlled trials demonstrated a significant reduction in state and trait anxiety and a significant increase in active coping skills among oncology nurses and female university students after using an unguided mobile app called Mobile Stress Management.
Substance use	
Use of Smartphone Technology	To date, there has only been one pilot feasibility study that uses a mobile application (DBT Coach) to reduce substance use (alcohol, drugs and tobacco) among adults suffering from borderline personality disorder. Using the application in addition to a face-to-face therapy session reduced emotional intensity and the need for substance use.

Current usage of technological advances in Psychiatry

Since 2006, journals such as Academic Psychiatry have proposed that knowledge and skills in medical informatics will be essential towards life-long learning and modern medical psychiatry practice. A needs based assessment was conducted, and it was previously proposed that existing psychiatry curriculum needed to integrate with the advancement in technologies. A pilot study done previously in 2006, looking at user perceptions of technology in psychiatry, demonstrated that residents and medical students in 2006 already felt that technology skills are integral in their medical training. As of 2006, residents and medical students have already indicated a preference towards usage of PDAs (personal digital assistants) as they help to provide immediate access to critical information that was pertinent towards the clinical care of patients as compared to the conventional modalities of checking for evidence based information. More recent research has looked into the implementation of E-learning in medical psychiatry education by the Medical Faculty of the University of Ulm. This particular feasibility trial showed that knowledge about a child psychiatry module could be similarly imparted via an online platform, as compared to previous more conventional modalities of learning. Even more recently, psychiatric educators at the Hamburg medical school have also made use of "Cinemeducation" to help students gain a deeper understanding of psychiatric illnesses.

Even more recently, a literature review was done using the keywords "Smartphone, Applications, Psychiatry". From our literature review search, there have not been rigorous studies to date looking into web-based E-learning technologies and also smartphone based technologies on the educational of medical students and residents. The latest advancement in technology is the application of unique tele-technologies to help augment resident's training in psychodynamic psychotherapy. There is indeed a paucity of existing research looking into the application of the latest web-based and in particular smartphone technologies in psychiatry.

In order to systematically analyze the current breath of peer-reviewed literature for education in psychiatry, and to summarize the key findings of current research, and to identify current knowledge gaps, it might be worthwhile to consider a Scoping Review. Scoping reviews have been done by Katherine et al. (2013.) recently to assess the usage of technology in delivering mental health service for children and youths in Canada. Descriptive numeric summary and thematic analysis were conducted for the literature reviewed. In summary, the authors in their review concluded that the usage of technology did indeed play a major role in the service and supports to children and youth, and also in terms of prevention, assessment, diagnosis and treatment. Similar scoping review should be adopted to assess systematically the application of technology for Psychiatry as a specialty, as it would be key to understand the current breath of research; and to understand the impact of current research and also to systematically identify the key gaps in education. As no such reviews have been conducted, the authors were limited to a literature review using Medline and Pub-med, considering only indexed articles and using the keywords "Psychiatry, Education, Technology, Smartphones" to search for recent advances in psychiatry education using technology.

References for Chapter 1 & 2:

Hamid A. & Kavit A (2011). Smartphone Applications for the Urology Trainee. BJU International 2011:108, 1371-1375.

Karl F.B.P, Heather W. & Kim W. (2012). Smartphone and medical related App use among medical students and junior doctors in the United Kingdom (UK): A regional Survey. BMC Medical Informatics and Decision Making, 2012, 12:121

Mobile Future. 2010. Social Media, Apps, and Data Growth Headline 2010 Mobile Trends URL:
http://www.mobilefuture.org/content/pages/mobile_year_in_review_2010?/yearendvideo [accessed 2014-03-17]

Free C, Phillips G, Felix L, Galli L, Patel V, Edwards P. The effectiveness of M-health technologies for improving health and health services: a systematic review protocol. BMC Res Notes 2010;3:250

Abu S.M.M, Illhoi Y., & Lincoln S. (2012). A Systematic Review of Healthcare Applications for Smartphones. BMC Medical Informatics and Decision Making 2012, 12:67

Hilty DM, Hales DJ, Briscoe G., et al. APA Summit on Medical Student Education Task Force on Informatics and Technology: learning about computers and applying computer technology to education and practice. Acad Psychiatry; 2006 Jan-Feb;30(1);29-35.

Briscoe GW, Fore Arcand LG, Lin T, et al. Students' and residents' perceptions regarding technology in medical training. Acad Psychiatry, 2006 Nov-Dec; 30(6);470-479.

O'Conner P, Byrne D. Butt M, et al. Interns and their smartphones: use for clinical practice. Postgrad Med J, 2014 Feb;90(1060):75-79.

Tim R., Thomas C, Haider I et al. (2013). Smartphone Use and Acceptability Among Clinical Medical Students: A Questionnaire-Based Study. J Med Syst (2013) 37:9936 Handyman W, Bullock A, Brown A et al. (2013). Mobile technology supporting trainee doctors' workplace learning and patient care: an evaluation. BMC Medical Education 2013, 13:6

Payne K.F.B, Weeks L., Dunning P. (2014). A Mixed Methods pilot study to investigate the impact of a hospital-specific iPhone application (iTreat) within a British junior doctor cohort. Health Informatics Journal 2014, Vol20(1) 59-73

Hawkes C.P., Walsh B.H., Ryan C.A., Dempsey E.M. (2013). Smartphone technology enhances newborn intubation knowledge and performance amongst paediatric trainees. Resuscitation 84(2013) 223-226.

Hassani T.J.R, Sanharawi E.M., Dupont-Monod S et al. (2013) Smartphones in ophthalmology. J Fr Ophtalmol 2013 Jun;36(6):499-525

Bhasin B, Estrella M.M, Choi MJ. (2013) Online CKD Education for Medical Students, Residents, and Fellows: Training in a New Era. The National Kidney Foundation, Inc.

Nada AH, Sudip G. (2013) Smartphones and the plastic surgeon. Journal of Plastic, Reconstructive and Aesthetic Surgery (2013) 66, e155-e166

Nawfal Ah, Panagiotis DG, Shafic SA (2012). Smartphones in orthopaedics. International Orthopaedics (2012) 36:1543-1547.

Orrin IF, Timonthy FT.(2012) Smartphone App use among medical providers in ACGME Training Programs. J Med Syst (2012) 36:3135-3139.

Faye H, Richard RWB, Simon M. (2013). Smartphone apps to support hospital prescribing and pharmacology education: a review of current provision. British Journal of Clinical Pharmacology 77:1, 31-38.

Subhi Y, Todsen T, Ringstd C, Konge L: Designing web-apps for smartphonescan be easy as making slideshow presentations. BMC Res Notes, 2014 Feb 20;7(1):94 Thomas LL (2013). A Systematic Self-Certification Model for Mobile Medical Apps. J Med Internet Res 2013; 15(4):e89.

Abu S.M.M, Illhoi Y., & Lincoln S. (2012). A Systematic Review of Healthcare Applications for Smartphones. BMC Medical Informatics and Decision Making 2012, 12:67

Ammenwerth E, Buchaurer A, Blaudau B, Haux RL Mobile information and communication toools in the hospital. International journal of medical informatics 2000, 57:21-40.

A.W. Buijink, B.J. Visser, and L.Marshall. Medical apps for smartphones: lack of eveidence undermines quality and safety. Evidence-based Medicine, vol. 18, 90-92, 2013.

Bsoul M, Minn H, Tamil L. Apnea MedAssist: real-time sleep apnea monitor using single-lead EEG. IEEE Trans Inf Technol Biomed 2011 May;15(3):416-427.

BinDhim NF, Sharman Am, Trevena L, Basyouni MH, Pont LG, Alhawassi TM. Depression screening via a smartphone application: cross-country user characteristics and feasibility. J Am Med Inform Assoc 2014 oct 17.

Charpentier G, Benhamou P.Y., Dardari D, Clergeot A, Franc S, Schaepelnck-Belicar P., Catargi B., Melki V., Chailous L., Farret A., Bosson J.L., Perfomis A: The Diabeo Software Enabling Individualized Insulin Dose adjustments combined with Telemedicine Support improves HbA1c in poorly Controlled Type 1 Diabetic Patients: A 6 month, randomized, open-label, parallel-group, multicenter trial (TeleDiab 1 Study). Diabetes Care 2011, 34:533-539.

Edgar S, Swyka T, Fulk G, Sazonov ES. Wearable shoe-based device for rehabilitation of stroke patients. Conf Proc IEEE Eng Med Biol Soc 2010;2010:3772-3775

Free C, Phillips G, Felix L, Galli L, Patel V, Edwards P. The effectiveness of M-health technologies for improving health and health services: a systematic review protocol. BMC Res Notes 2010;3:250

Faurholt-Jepsen M, Frost M, Vinberg M, Christensen EM, Bardram JE, Kessing LV. Smartphone data as objective measures of bipolar disorder symptons. Psychiatry Res. 2014 Jun 30;217(1-2):124-127.

Granholm E, Loh C, Swendse J: Feasibility and validity of computerized ecological momentary assessment in schizophrenia. Schizophr Bull 2008, 34:507-514.

Granholm E, Ben-Zeev D, Link PC, Bradshw KR, Holden JL: Mobile assessemnt and treatment for Schizophrenia: A pilot trial of an interactive text-messaging intervention for medication adherence, socialization and auditory hallucinations. Schizophr Bull 2011, 38(3):414-425.

Grunerbl A et al. Smart-Phone Based Recognition of States and State Changes in Bipolar disorder patients. IEEE J Biomed Health Inform 2014 Jul 25 [Epub ahead of print].

Huang CC, Lee PY, Chen PY, Liu TY. Design and implementation of a smartphone-based portable ultrasound pulsed-wave Doppler device for blood flow measurement. IEEE Trans Ultrason Ferroelectr Freq Control 2012 Jan;59(1):182-189.

Palmier-Claus et al: The feasibility and validity of ambulatory self-report of psychotic symptoms using a smartphone software application. BMC Psychiatry 2012; 12:172.

Palmier-Claus et al: Integrating mobile-phone based assessment for psychosis into people's everyday lives and clinical care: a qualitative study. BMC Psychiatry 2013: 13:34.

Karl F.B.P, Heather W. & Kim W. (2012). Smartphone and medical related App use among medical students and junior doctors in the United Kingdom (UK): A regional Survey. BMC Medical Informatics and Decision Making, 2012, 12:121

Marhsall A, Medvedev O, Antonov A: Use of smartphone for improved self-management of pulmonary rehabilitation. International Journal of telemedicine and applications 2008.

Mobile Future. 2010. Social Media, Apps, and Data Growth Headline 2010 Mobile Trends URL: http://www.mobilefuture.org/content/pages/mobile_year_in_review_2010?/yearendvideo [accessed 2014-03-17]

Meankaew P, Kaewkungwal J, Kharmsiriwathara A, Khunthong P, Singhasivanon P, Satimai W: Application of mobile-technooyg for disease and treatment monitoring of malaria in the "Better Border Healthcare Programme". Malar J 2010;9:237.

Ozdalga E., Ozdalga A., Ahuja N. The Smartphone in Medicine: A Review of Current and Potential use Among Physicians and Students. J Med Internet Res 2012; 14(5):e128.

Oresko JJ, Duschl H, Cheng AC. A wearable smartphone-based platform for real-time cardiovascular disease detection via electrocardiogram processing. IEEE Trans Inf Technol Biomed 2010 May;14(3):734-740.

Martinez-Perez B, da la Toree-Diez I, Loprez-Coronado M. Mobile Health Aplications for the Most Prevalent Conditions by the World Health Organization: A Review and Analysis. J Med Internet Res 2013;15(6):e120.

Subhi Y, Todsen T, Ringstd C, Konge L: Designing web-apps for smartphonescan be easy as making slideshow presentations. BMC Res Notes, 2014 Feb 20;7(1):94
Thomas LL (2013). A Systematic Self-Certification Model for Mobile Medical Apps. J Med Internet Res 2013; 15(4):e89.

Spaniel F, Vohldka P, Kozeny J, Novak T, Hrdicka J, Motlova L, Cermak J, Hoschl C: The information Technology Aided Relapse Prevention Programme in Schizophrenia: an extension of mirror-design follow-up. Int J Clin Pract 2008, 62:1943-1946.

Rajput Za, Mbugua S, Amadi D, Chepngeno V, Saleem JJ, Anokwa Y, et al. Evaluation of an Android-based mHealth system for population surveillence in developing countries. J Am Med Inform Assoc 2012 Jul 1;19(4):655-659.

Sposaro F, Danielson J, Tyson G. iWander: An android application for dementia patients. Conf Proc IEEE Eng Med Biol Soc 2010;2010:3875-3878

Wu HH., Lemarie E.D., Baddour N. Change –of-state determination to recognize mobility activity using a Blackberry smarthone. Conf Proc IEEE Eng Med Bio Soc 2011; 2011:5252-5255.

Worringham C, Rojeck A, Stewart I. Development and feasibilty of a smartphone, ECG and GPS system for remotely monitring exercise in cardiac rehabilitation. PLoS One 2011;6(2):e14669.

Chapter 3: Application of Internet and Smartphone Technologies for education in Psychiatry

The global outreach of a locally developed Psychiatry Application for Undergraduate Psychiatry Education

Introduction

Over the past decade, there have been massive developments in both web-based and smartphone technologies. It was perhaps the release of Apple's iPhone in 2007 and the launch of the Apple Application store in July 2008 that was pivotal in causing a huge change in the way the world uses the web. Smartphones are considered to be a new generation of mobile technology, that are equipped with immense computing capabilities that allows individuals to have constant access to the internet and make use of applications for various usages [1].

Previous studies have looked into medical students and trainee's ownership and perspectives towards smartphone usage. In 2012, a questionnaire-based survey was distributed amongst the interns in the Republic of Ireland [2]. The survey demonstrated that smartphone were widely adopted and accepted and that they were being used by interns to aid them in performing their daily duties. The survey highlighted that the most popular application that was commonly used was that of the British National Formulary application [2]. Other studies have highlighted similar results, in that there was a high usage rate of smartphone and its applications amongst medical students and junior doctors and other studies have also found that iPhone users tend to own more applications [3]. Previous studies indicate that students and junior doctors make use of application for an estimated 30 minutes each day [3]. Given the results of the previous questionnaire surveys done, it is obvious that smartphone technologies are well accepted by students, trainees and doctors. It would be of interest to determine how several specialties have made use of a hybrid of online and smartphone technology in educational settings.

A review of the current literature on the following database, which were searched from inception to July 2014: Pubmed (from 1966), EMbase (from 1980), PsychINFO (from 1806), BIOSIS (from 1926), Science Direct (from 2006) and Cochrane CENTRAL (from 1993) using the search terminologies: education, smartphone have identified the following specialties to have used smartphone technologies in education: Pediatrics, Ophthalmology, Nephrology, Plastic Surgery, Orthopedics, Pharmacology and Urology. For example, in Pediatrics, a smartphone neonatal intubation application was deployed to enhance overall intubation skills [4]. In Ophthalmology, a cumulative total of 342 applications have been identified to be of value in terms of enhancing clinical skills [5]. In Nephrology, several online web based resources were identified to be of value for medical students and residents to augment their knowledge with regards to the complications of chronic kidney disorders [6]. In Plastic Surgery, 16 applications that are of educational value to the plastic surgeon have been identified [7]. It would thus be of interest for the authors to determine to what extent Psychiatry, as a discipline has embraced online, web-based and smartphone technologies for educational needs of psychiatry medical students and residents. A search into the literature revealed that the most recent application of web-based technology was that of the usage of virtual worlds for role-play simulation in child and adolescent psychiatry [8] and the usage of telemedicine for peer-to peer psychiatry learning between medical students in the United Kingdom and Somaliland [9]. A search through the existing published literature using the keywords "psychiatry, smartphone, education" did not yield any published papers to date that examined the application of latest web-based and smartphone technologies for psychiatry education.

The authors hope to make use of the latest web-based and smartphone technologies in implementing both a web-based as well as a native smartphone psychiatry textbook companion for undergraduate students in Psychiatry.

The main objectives of the current paper and research are as follows: *To determine the feasibility and acceptance of a locally developed Psychiatry Smartphone application and initial user perspectives towards it; and to determine the feasibility and acceptance of a locally developed Psychiatry Smartphone application globally for Psychiatry education internationally.*

Methodology
The Mastering Psychiatry Smartphone Application (Web-application & Native application)

A newly written textbook (jointly written by both the authors) that integrated both localized (Singapore) and United Kingdom Clinical guidelines were initially crafted in 2011. The authors filmed videos demonstrating assessment methodologies for the various psychiatric disorders locally in Singapore. The core textbook materials as well as the videos were then integrated into a web-based smartphone application. The web-based smartphone application was programmed using an online application builder and a blogging website using both HTML5 and Java-script coding. Videos were stored online on a video-hosting website (www.vimeo.com) and embedded within the web-based smartphone application. In addition, a questionnaire based quiz that contains both multiple choice questions and short answered questions was crafted using an online questionnaire builder and integrated into the web-based smartphone application. Prior to the launch of the web-based smartphone application, the usability of the application was evaluated across several different computing platforms to ensure that the system was robust.

In 2013, the authors, MWBZ and RCMH were offered an educational grant for the development of a native Apple and Android Smartphone application. The educational grant, offered by Provost Office, National University of Singapore was catered for innovations that could enhance current learning methodologies. The native application that was eventually developed by an external vendor contained largely the same contents as the web-based application developed by the authors, with the exception that there was a text messaging event management system integrated to enable students to schedule tutorials and receive text reminders for tutorials. The Apple IOS sand Android Play applications were made available on the app store in the late 2013.

Information about the application was disseminated via printed materials locally in Singapore, as well as by means of a short demonstration to the undergraduate students at the National University of Singapore, Yong Loo Lin School of Medicine on the first day of their clinical psychiatry posting. With ethics approval from the National University of Singapore, a user perspective survey was administered to our students immediately after the end of their end of posting test to determine user's perspectives towards the usefulness of such an application.

Results

The web-based smartphone application (www.masteringpsychiatry.com) was launched on July 15, 212 and from the inception of the application till the time of preparation of the manuscript, there has been a cumulative total of 28,500 unique visits of the responsive HTML5 web-based smartphone application. The majority of the users are from Singapore, with 19,383 total views from Singapore. The other 2 countries that ranked highest were that of the United States with 1,487 unique views as well as Malaysia with 942 unique views. Figure 1 show the geographical utilization of the portal since inception till the time of preparation of this manuscript.

------------------------ Insert Figure 1 --

Country	Views
Singapore	19,385
United States	1,487
Malaysia	942
India	785
Australia	752
United Kingdom	677
Hong Kong	272
Turkey	258
Canada	251
Saudi Arabia	248
Pakistan	234
Nigeria	160
Philippines	150
Ireland	141
Egypt	125
New Zealand	94
South Africa	89
Indonesia	73
Yemen	70
Kuwait	67
Oman	67
Jamaica	64
Germany	64
United Arab Emirates	64
Macao	63

Figure 1: Geographical Map of Utilization of the Web-based application since inception

With regards to the utilization of the native applications, from the inception of the applications on the respective application stores to date, there have been a cumulative total downloads of 2200 downloads of the Mastering Psychiatry application from the Apple app store and 7000 downloads of the same application from the Android app store. Figure 2 shows the cumulative total number of downloads from each of the respective application stores. From our knowledge, the native application has been granted a 4+ out of 5 score on the Apple Application store, whilst on the Android application store, it has been granted a 4.5 out of 5 score. A total of 161 users have rated our application on the Android store and a cumulative total of 143 out of the 161 users have given the application a score of 3 or more. Some of the qualitative feedback made available on the android store was that the application was deemed to be a great book for beginners and that it was an excellent application. Some users did bring to the authors and the developers attention technical issues pertaining to the usage of the application on some devices.

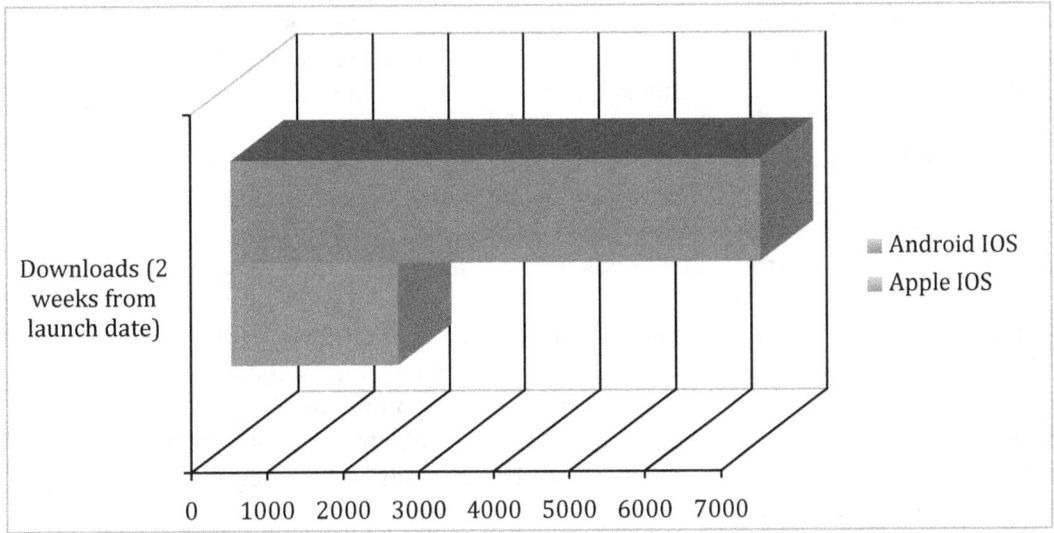

Figure 2: Cumulative total number of downloads from each of the respective application stores.

Locally, in Singapore, a total of 185 students took part in the user perspective survey. The majority of them were of the age of 22 years old (79.2%) and the majority of the students used an Apple IOS device (53.3%). The initial user perspective survey conducted highlighted that approximately a total of 95.2% of students felt that having a psychiatry smartphone application was deemed to be useful. 71.4% of the students agreed that a smartphone application would be a good added companion to a traditional textbook.

Discussion
From the best knowledge of the authors, this study is perhaps one of the first studies to demonstrate the success of the implementation of smartphone technologies for educational needs in Psychiatry. The initial findings demonstrated the feasibility and the acceptance of smartphone applications for Psychiatry in Singapore, as well as the feasibility and acceptance of smartphone applications for Psychiatry globally. In addition, the respective app ratings and the results obtained from the initial user perspective survey showed that the design of the application was appropriate and deemed useful for students, and that the application has a role in augmenting current educational needs of students on the go.

The usefulness of smartphone applications for education has been supported by prior research. Tripathi et al. (2014) [10] have conducted a review of relevant applications for neurosurgery on the respective application stores and highlighted that students and medical doctors prefer applications that have links to scoring systems, operative illustrations as well as textbook based contents. In consideration of the previous findings, the authors postulate that the current success of our application internationally might be because of the fact that our current application offers students not only access to textbook based materials, but also access to other materials such as videos and questionnaire quizzes. These might be relevant and deemed useful with regards to helping students to acquire the necessary knowledge in Psychiatry.

Perhaps, the closest correlation with our current study is that of the previous study done by Waldmann UM and Weckbecker L (2013) [11]. They have similarly designed a web-based application to help teach their medical students about primary care guidelines. Their results noted that amongst their group of 14 student testers, it was noted that the majority of the students made use of the application more frequently, and also made use of their free time to study the guidelines. The authors concluded that their self-designed smartphone application has helped to create interest amongst student and has helped them to acquire valuable knowledge. Similarly, our self-created web-based and native application has enabled our students to learn on the go, as well as help to augment their learning needs in psychiatry.

The main strength of the current study is that the authors managed to demonstrate the feasibility of implementation of a smartphone application for psychiatry locally and internationally. The current study has also managed to demonstrate that local Singaporean students perceive smartphone applications in Psychiatry positively. Our current studies have replicated some of the findings of other studies with regards to the usage of applications for education.

Nevertheless, there are several limitations to the current study. We acknowledge that whilst there is a good number of viewership of our web-based application, we do not have the absolute number of individual users who have downloaded the web-based application, as we are limited by the database being able to only track individual unique access. This information, in conjunction with the platform that users view the application is crucial in terms of designing future educational applications, as well as planning future studies. We are also limited by the fact that we could not acquire user's perspectives from the Apple app store. In addition, our perspective survey has only been administered to our local cohort of students and might not be entirely representative of the views of the global audience. To mitigate this limitation, we would need to find liked-minded individuals from organizations overseas to collaborate and perform a comparative study with regards to user perception of our educational application.

Conclusions

This study is perhaps one of the first to demonstrate the feasibility and the global acceptance of a locally self-designed educational application for psychiatry education. Whilst the current research has managed to demonstrate the feasibility and acceptance of such an application, future studies would be warranted to look in-depth into whether there are difference culturally in terms of perceptions towards having such an application in psychiatry and what contents different culture and cohort of student might want within an application.

References:
1. Hamid A. & Kavit A (2011). Smartphone Applications for the Urology Trainee. BJU International 2011:108, 1371-1375
2. O'Conner P, Byrne D. Butt M, et al. Interns and their smartphones: use for clinical practice. Postgrad Med J, 2014 Feb;90(1060):75-79.

3. Payne K.F.B, Weeks L., Dunning P. (2014). A Mixed Methods pilot study to investigate the impact of a hospital-specific iPhone application (iTreat) within a British junior doctor cohort. Health Informatics Journal 2014, Vol20(1) 59-73

4. Hawkes C.P., Walsh B.H., Ryan C.A., Dempsey E.M. (2013). Smartphone technology enhances newborn intubation knowledge and performance amongst paediatric trainees. Resuscitation 84(2013) 223-226.

5. Hassani T.J.R, Sanharawi E.M., Dupont-Monod S et al. (2013) Smartphones in ophthalmology. J Fr Ophtalmol 2013 Jun;36(6):499-525

6. Bhasin B, Estrella M.M, Choi MJ. (2013) Online CKD Education for Medical Students, Residents, and Fellows: Training in a New Era. The National Kidney Foundation, Inc.

7. Nada AH, Sudip G. (2013) Smartphones and the plastic surgeon. Journal of Plastic, Reconstructive and Aesthetic Surgery (2013) 66, e155-e166

8. Vallence AK, Hemani A, Fernandez V, Livingstone D, McCusker K, Toro-Troconis M. Using virtual worlds for role play simulation in child and adolescent psychiatry: an evaluation study. Psychiatry Bull (2014) Oct; 38(5):204-10.

9. Keynejad R, Ali FR, Finlayson AE, Handuleh J, Adam G, Bowen JS, Leather A, Little SK, Whitwell S. Telemedicine for pee to peer psychiatry learning between UK and Somaliland medical students.

10. Tripathi M, Deo RC, Srivastav V, Baby B, Singh R, Damodaran N, Suri A. Neurosurgery apps: novel knowledge boosters. Turk Neurosurg. 2014;24(6):828-38.

11. Waldmann UM, Weckbecker K. Smartphone application of primary care guidelines used in education of medical students. GMS Z Med Ausbid 2013;3(1):Doc 6.

Chapter 4: Application of Video technologies and Augmented reality Technologies for education in Psychiatry

Perceptions towards Video Technologies and Augmented Reality in Psychiatry Education

Introduction

Since 2006, academic journals in Psychiatry have already proposed that knowledge and skills in medical informatics will certainly be essential and crucial towards life-long learning in psychiatry and also in modern day psychiatry undergraduate and postgraduate examination. A previous needs based assessment was conducted in 2006, which highlighted that the existing psychiatry curriculum in 2006 needed to integrate and make use of the advancement in technologies [1]. A subsequent pilot study was conducted, looking at user perspectives towards technology in Psychiatry, and it demonstrated that resident and medical students, back in 2006, have already felt that technology skills were essential in their medical training [2]. At that time, resident and medical students already felt that PDAs (personal digital assistants) were able to provide them with immediate access to evidence based information that could help them in management of their patients [2]. Over the past decade, there have been major advances in terms of Web 2.0 technologies as well as in terms of smartphone-based technologies that will definitely have an impact on the way undergraduate and postgraduate students learn.

In psychiatry, the most recent application of the advances in Web 2.0 technologies and smartphone technologies has been in the usage of these tele-technologies in augmenting resident's training in psychodynamic therapy [3]. A recent narrative review conducted [4] has identified 20 previous publications relating to tele-psychiatry in education, but have cautioned against the evidence base pertaining to the identified literature, as most of the published studies are descriptive in nature and the methodologies are heterogeneous as well [4]. Psychiatry, has also adopted the usage of other innovative forms of technologies such as smartphone technologies and virtual reality. Zhang MW et al. (2014) [5] recently described the methodology of development of a web-based psychiatry online portal and a smartphone application, and their initial quantitative analysis has demonstrated that Asian students were receptive towards the newer innovations in technologies. Virtual reality technologies have also been recently evaluated for psychiatry education. Vallence et al. (2014) [6] developed and evaluated a novel teaching session, using a computer program (Second life) to conduct clinical assessment using role-play stimulation. Participants reported much less anxiety in the virtual setting as compared to role-playing face to face, and they did report similar improvements in psychiatric skills and knowledge. However, the limitation with the recent study was with the low sample size of 10 students.

Apart from the interest in utilizing Web 2.0 and smartphone technologies in education for psychiatry, several other disciplines have described the added efficacy of augmentation of students' education needs using video technologies as well as other modalities of technologies, such as augmented reality. Hibbert EJ et al. (2013) [7] previously described a randomized controlled pilot trial that compared the impact of access to clinical endocrinology video demonstrations with access to usual resources on medical students performance of clinical endocrinology skills. Their pilot randomized study have determined that exposure to high quality videos demonstrating clinical skills could significantly improve performance in observed structural clinical examinations of such skills. They have recommended the usage of such technologies for education as it is considered to be cost-effective for large numbers of learners. Grynberg M et al. (2012) [8] have compared student's perception towards learning pelvic and breast examinations using a training model against videos. In their research, which sampled a total of 79 2nd and 3rd year undergraduate students, it was demonstrated that there is a higher degree of satisfaction amongst the students for the video clip as compared to the training model. Augmented reality is another recent learning methodology, as it is capable of providing learning opportunities in which virtual learning experiences could be inter-linked and integrated in a real-life physical context. With regards to the utility of augmented reality in healthcare education, Zhu E et al. (2014) [9] have published an integrative review of augmented reality for education. Zhu E et al. (2014) [9] have done a thematic analysis based on 3 qualitative, 20 quantitative and 2 mixed studies, to which they report that augmented reality could indeed be applied to a variety of topics in medical and healthcare education. Zhu E at al. (2014) [9] have also concluded that learners are more willing to accept this form of technology and augmented reality has in itself, the potential to improve a diverse range of clinical competencies. However, in the recent integrative review published, the authors did cautioned about the fact that most of the current studies done recently are still early pilot studies and lack a well-designed framework.

Previously, it has been highlighted that in the recent decade, tele-technologies, smartphone technologies and even virtual reality technologies have been utilized for psychiatry education. These newer modalities of technologies have demonstrated that they could augment clinical teaching. Given the current reported efficacy of video as well as augmented reality technologies in other disciplines, further studies using these methodologies in psychiatry might be warranted, given that there is currently no studies looking into the educational efficacy of videos and augmented reality specifically for psychiatry. With this in mind, the authors have applied a simple developmental process of building an online portal hosting clinical psychiatry videos as well as implementation of augmented reality in psychiatry, as described below in the methodologies section. Our research objectives were: *(a) to determine whether undergraduates students will be receptive towards clinical psychiatry videos as an added aid for learning and (b) to determine whether students will be receptive towards augmented reality technologies in enhancing their learning experience in psychiatry.*

Methodology
The Mastering Psychiatry Online Videos Vault

The Mastering Psychiatry Online Videos Vault was developed between February 2012 and June 2012. The augmented reality features was developed in September 2014 and launched since then.

The authors made use of digital video technologies and have filmed the following videos, demonstrating to students how to assess patients for a particular psychiatric disorder, and how to elicit basic psychopathologies. These videos include that of:

1. Psychosis: History taking

2. Depression: History taking
3. Anxiety: History taking
4. Explanation of antidepressants
5. Explanation of Cognitive behavioral therapy
6. Assessment of Borderline Personality disorder
7. Suicide risk assessment
8. Explanation of Electro-convulsive therapy treatment
9. Frontal Lobe Examination
10. Mini Mental State examination
11. Explanation of Lithium therapy
12. Explanation of Neuroleptic Malignant Syndrome
13. Violence Risk Assessment
14. Explanation of Dementia Medications
15. Sleep disorder assessment

Developmental Process
The developmental process of setting up the online video vault was undertaken by the author MWBZ. The online portal was build using a web portal designer that utilized graphic user interface (www.wordpres.com). The videos that were filmed were hosted online on a web-based video storage service (www.vimeo.com) and the direct links to the videos were incorporated into the online portal. A free augmented reality software was used to incorporate augmented reality elements into the existing hardcopy textbook, which was made freely available for students to download on the web-portal and print. The elements and functionality of augmented reality incorporated include that of (a) live links to the web-based portal, (b) live links to the online assessment quiz, (c) live link to send enquiries to our centralized email service and (d) live links to videos hosted online when students flip through certain pages of the existing hardcopy textbook. The usability of the online video portal as well as the augmented reality feature were evaluated by the authors across several computing platforms to ensure the robustness of the system. Figure 1 show the augmented reality features integrated into the existing print materials.

-------------------------- Insert Figure 1 -----------------------------

Figure 1: Augmented reality features (as visualized using the Smartphone Application 'Layar') incorporated onto the front, back cover & Objective Structured Clinical Examination Interview Grids within the book.

<u>Dissemination and collation of perspectives</u>
With regards to the deployment of the videos online, all of the students who were posted to the National University of Singapore, Yong Loo Lin School of Medicine Psychiatry Undergraduate program, were provided with information about the portal on the first day of their clinical posting. A ten-minutes demonstration of the online portal as well as the augmented reality features was conducted by MWBZ. RH is the undergraduate education director, and was involved in allocating time for MWBZ to demonstrate the features online.

With ethics approval from the National University of Singapore, a user perspective survey was administered to the students, right after the completion of their end of posting clinical assessment. Participation in the survey is entirely voluntary and relevant participant information handout was provided to all the participants prior to the start of the survey. The user perspective survey looks specifically into student's perception of the features made available to them. The questions asked involved the themes as to whether students did perceive certain contents to be relevant and helpful for psychiatry education.

Statistical analysis was performed using SPSS version 16.0. The differences between the male and female gender in terms of their perspectives were analyzed using the chi-squared test.

Results
Since the introduction of the online video portal to the undergraduate students till the end of their final examinations in January 2014, there has been a cumulative total of 58,635 independent views of the videos online. Figure 2 shows the average number of views per month.

-------------------- Insert Figure 2 --

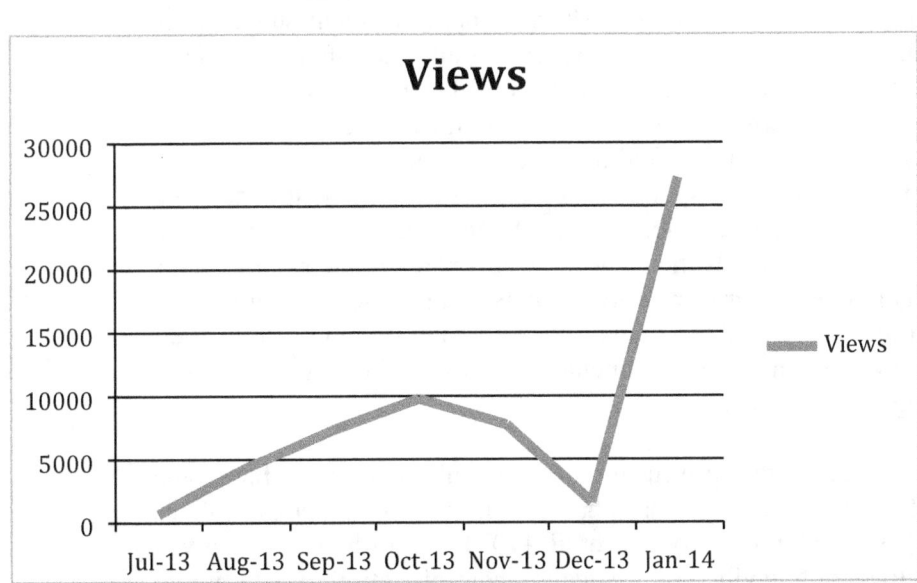

Figure 2: Line-graph showing the Average cumulative number of views of the online videos per month.

The augmented reality features were only developed in September 2014, and only launched to the current batch of undergraduate students doing psychiatry (n=60). A cumulative total of 47 individuals have accessed and made use of the interactive elements on the front and back cover. Another 50 individuals have accessed and made use of at least 1 out of the 6 objective structured clinical examination (OSCE) grids in the hard copy textbook. This amounts to a response rate of 78.3% for the interactive elements on the cover pages and a response rate of 83.3% for the interactive elements on the OSCE grids in the book.

A cumulative total of 185 students took part voluntarily in the user's perspective survey. The majority of the undergraduate students were of the year of 22 years old (79.2%) and most of them were using an Apple IOS device (53.3%). With regards to ownership of smartphone applications, the vast majority of students (66.7%) had between 1 and 5 medical applications loaded on their smartphones. The usage of the medical applications was mainly for augmentation of their educational needs. Most students tend to use their smartphone devices either in the clinics (20.4%) or the ward (31.1%).

Out of the total number of students, the vast majority find that having clinical OSCE videos to help augment their educational needs to be useful (43.3%) and an additional 23.0% of students have deemed clinical videos to be extremely useful for learning. Further statistical analysis conducted revealed there is also no noted difference between gender and student's perception of having clinical OSCE videos (χ^2= 1.278, p=0.865).

Discussion
From our current knowledge, this is perhaps one of the first few studies to describe and evaluate student's perceptions toward incorporating clinical videos to augment psychiatry teaching. In addition, this is one of the first studies to implement and evaluate the initial receptiveness towards the utilization of augmented reality technologies in psychiatry. Our current findings showed that our Asian students are receptive towards the implementation of clinical videos to augment their psychiatry educational needs; and that a significant high proportion of the students actually have had a positive perception with regards to the inclusion of clinical videos to meet their psychiatry educational needs. Our current findings show that there is a peak in the usage of the clinical videos, which corresponds to the months to which students' end of posting examinations were conducted (September, October and November), and a higher peak during the month to which their end of year professional examinations were conducted (January). The response rate to the newly implemented augmented reality features was around 78% to 83%, which showed that students doing their undergraduate psychiatry posting were amenable towards trying out new technologies, such as augmented reality.

Our current findings with regards to the utilization of clinical videos for educational needs is in keeping with what previous research has determined, for other disciplines such as endocrinology [7]. In addition, Holland et al. [10] have in their study, given nurses unlimited access to an online video of medication administration, in addition to standardized lectures and skills classes typically used to teach this skill. Holland et al. [10] found that the inclusion of videos, and giving nurses unlimited access enabled them to perform better on observed structural clinical assessment on medication administration, and in addition, it also helped in improving their satisfaction with teaching. Our current video vault works operates in a similar methodology, in which students have unlimited access to the video vault, right from the first day of their clinical posting. This might have, in similar ways as the Holland et al. [10] study, helped in the improvement of their clinical skills, and might have made them more satisfied with the quality of the teaching methodologies. The positive perceptions that our students have towards inclusion of videos into the clinical psychiatry curriculum could are also in keeping with the previous findings of Kohgali SEO et al. [11] and Roshier AL et al. [12]. Kohgali SEO et al [11] found that 96% of the students at the University of Dundee rated e-learning resources introduced to their cardiology curriculum as having probably or definitely of value. A high proportion of the students found the animations, the self-assessment exercises and video demonstrations of value. Roshier AL at al. [12] found that veterinary students at the University of Nottingham perceive online videos as being able to enhance teaching, as they are more accessible.

In addition, our initial findings with regards to our student's receptiveness towards newer modalities of technology is also in keeping with what has been previously found by other studies. Ponce BA et al. (2014) [13] investigated the use of augmented reality in orthopaedic education, and their pilot study not only revealed the safety of implementation of such a technology in their discipline, but also highlighted that, residents and surgeons concurred that their training was enhanced.

The main strength of the current study is that we managed to determine the educational effectiveness of video technologies as well as determining the receptiveness of psychiatry students towards newer modalities of technologies such as augmented reality. Our current study has empowered students, and possibly trainees and residents to have an opportunity to make use of the latest innovations in technologies. In addition, the initial results also demonstrate the feasibility of adopting this methodology for other psychiatry educational needs, either for undergraduates or post-graduates. Nevertheless, there remains to be several limitations in our current study. Our sample size survey is small, and only comprised of an Asian cohort of students. There is a lack of further evaluation of student's perceptions towards augmented reality, as it is a relatively new technology that has just been implemented this year. The authors acknowledged that there are limitations pertaining to the assessment of usability, as the statistics captured online reflect the total number of visits to the site by all visitors; which in turn implies that the authors were not able to identify the absolute number of unique visitors to the site. Diversity in terms of the opinions of the students could not be assessed, as focus group analysis was not conducted. It is tough to identify students who have not used the on-line resources, as pertinent information about the online resources was disseminated since the first day of their clinical attachment, and hence it is technically difficult to compare the differences in perception between users and non-users.

Conclusions

This is one of the initial studies that have demonstrated the effectiveness of including videos to augment psychiatry education, and one of the first studies to determine student's perception of using augmented reality in psychiatry. It is hoped that more clinicians will be willing to consider using the above methodologies and that there will be further studies done to evaluate changes in satisfaction scores as well as objective improvement in clinical skills. Our methods might apply to future research involving the use of technology, not only in undergraduate education, but also potentially for postgraduate education. Our current study demonstrated the feasibility of such technologies in education and has also demonstrated that students perceive videos as a useful addition to their educational needs and are receptive towards augmented reality, to augment learning needs. These have been proven in previous research in other medical and surgical domains and have been replicated in our current study.

References:
1. Hilty DM, Hales DJ, Briscoe G., et al. APA Summit on Medical Student Education Task Force on Informatics and Technology: learning about computers and applying computer technology to education and practice. Acad Psychiatry; 2006 Jan-Feb;30(1);29-35.

2. Briscoe GW, Fore Arcand LG, Lin T, et al. Students' and residents' perceptions regarding technology in medical training. Acad Psychiatry, 2006 Nov-Dec; 30(6);470-479.

3. Katzman J, Abass A, Coughlin P, et al. Building connections through teletechnologies to augment resident training in psychodynamic psychotherapy. Acad Psychiatry: 2014 Mar 25.

4. Sunderji N, Crawford A, Jovanovic M. (2014). Telepsychiatry in Graduate Medical Education: A Narrative Review. Acad Psychiatry 2014 Aug 26 [Epub ahead of print].

5. Zhang MW, Ho CS, Ho RC (2014). Methodology of development and students' perceptions of a psychiatry educational smartphone application. Technol Health Care 2014 Oct 15 [Epub ahead of print].

6. Vallance AK, Hemani A, Fernandez V, Livingston D, McCusker K, Toro-Troconis M. (2014). Using virtual worlds for role play simulation in child and adolescent psychiatry: an evaluation study. Psychiatric Bulletin 2014;38:204-210.

7. Hibbert EJ, Lambert T, Carter JN, Learoyd D, Twigg S, Clarke S (2013). A randomized controlled pilot trial comparing impact of access to clinical endocrinology video demonstrations with access to usual revision resources on medical student performance of clinical endocrinology skills. BMC Medical Education 2013, 13:135.

8. Grynberg M, Thubert T, Guilbaud L, Cordier AG, Nedellec S, Lamazou F, Deffieux X. (2012). Student's view on the impact of two pedagogical tools for the teaching of breast and pelvic examination techniques (video-clip and training model): a comparative study. European Journal of Obstetrics and Gynecology and Reproductive Biology 164; (2012): 205-210.

9. Zhu E, Hadadgar A, Masiello I, Zary N. (2014). Augmented reality in healthcare education: an integrative review. PeerJ 2014 Jul 8;2:r469.

10. Holland A, Smith F, McCrossan G, Adamson E, Watt S, Penny K. 2012. Online video in clinical skills education of oral medication administration for undergraduate student nurses: a mixed methods, prospective cohort study. Nurse Edu Today 2012; 33:663-670.

11. Khogali SEO, Davies DA, Donnan PT, Gray A, Harden RM, McDonald J, Pippard MJ, Pringle SD, Yu N. 2011. Integration of e-learning resources into a medical school curriculum. Med Teach 2011, 33:311-318.

12. Roshier AL, Foster N, Jones MA. (2011). Veterinary students' usage and perception of video teaching resources. BMC Med Education 2011, 11:1.

13. Ponce BA, Jennings JK, Clay TB, May MB, Huisingh C, Sheppard ED. (2014). Telementoring: use of augmented reality in orthopaedic education: AAOS exhibit selection. J Bone Joint Surg Am. 2014 May 21;96(10):e84.

Chapter 5: Application of Internet and Smartphone Technologies for patient care

Enhancing Patient Care amongst GPs

Introduction

Mobile phones have proliferated over the last decade. In this decade, smartphones have perhaps emerged as the most revolutionary invention. Smartphones equipped with immensely advanced computing capabilities, allow individual users to access the Internet "on the go". In addition, with their high resolution display screens, advanced optics and well-developed operating systems, smartphones help individual users to manage some aspects of their daily life. With the improvement in smartphones related technologies, there has also been a massive surge in the number of smartphone applications being developed and made available for downloading. Statistics have also shown a tremendous increase in the number of downloads of smartphone applications, from 300 million in 2009 to over five billions in 2010 [1]. It is now increasing convenient for an individual to browse through the online application stores to find relevant applications for social interaction, entertainment, or education purposes [1].

With the pervasiveness of smartphones worldwide, it is definitely likely to have an impact on the healthcare industrial. Previous research studies have also demonstrated the effectiveness of smartphones in healthcare related research - mainly in collection and analysis of patient's data [2] With the capacity of geo-positioning (GPS), the use of smartphones to support the implementations of healthcare programs in rural and remote areas, where clinicians could get immediate access to core clinical information has been demonstrated. In addition, studies have shown that GPS features would help in monitoring of a whole series of behaviors, and such technologies have been useful in smoking research and treatment [3]. Moreover, other potential applications of smartphones in healthcare, such as the use of short messaging services to promote desired behavioral change, as well as to improve patient's compliance to medications and various treatment regiment have also been reported [4]. As of 2011, it has been estimated that around 7000 smartphones health related applications are currently available for users. An extensive review of the media for delivering health care has identified the superiority of mobile healthcare applications over traditional information and communication technologies [5]. Previous studies have looked into the impact that mobile phone have had on healthcare and also helped to identify the main areas of health care delivery to which mobile devices could have an impact - which was identified to be health related outcomes [6] This superiority has been mainly attributed to a combination of factors, including the portable and continuous online status available for smart-phones which enable the integration of various behavioral interventions into everyday life. More importantly, applications are able to provide cheaper, more convenient and less stigmatizing interventions. The superiority is also due to the ability of smartphones to feedback the information gathered to relevant health professionals, along with data acquired from the internal sensors that could provide valuable information about health related behaviors.

Although the number of healthcare related applications has been increasingly steadily over recent years, the majority are currently limited to providing information, advice, instruction, support, encouragement and to various interactive tools for individuals for monitoring, recording and reflection. Based on previous surveys [7] most of the current healthcare applications are targeted at the general public, and not the medical specialist or the primary healthcare provider. It is also of no doubt that healthcare applications have far lesser outreach in certain medical domains, especially at the primary care level. We have identified mental health and psychiatry as particular areas for potential development. There have been various healthcare applications written to provide basic information about mental health disorders and to track the emotion and mood of users, such as Epocrates, Psych Facts and Depression Monitor on the Apple i-Tunes store. Because the current applications largely target the general public, the common limitations in psychiatry are that they fail to provide primary providers with an integrated application that incorporates assessment tools and medical information targeting mainly for physicians. Moreover, they may not provide self-report questionnaires for patients during consultations, to enable physicians to track patient's individualized progress. *As a result, our we have proposed this mental health application primarily catering to the needs of primary care providers, in order to help keep primary healthcare providers updated with latest psychiatry knowledge and also to advocate earlier detection and interventions for at risk individuals in the general population.*

Methodology
The Mental Health Wellness smartphone application (MHWell App)
The web-based smartphone application was developed between January 2012 and April 2012, based on a developmental approach involving five phases: a) formulation of user requirements, b) system design, c) system development, d) system evaluation and e) system application. The details pertaining to each phase of system development are documented below.

a. Formulation of user requirements
The Mental Health Wellness (MHWell) App was developed in consideration of the current limitations in the field of mental health and psychiatry catering for primary care providers. The application aims to provide mental health practitioners with basic tools to screen for core mental health disorders, and to provide them with structured questionnaires for patients prior to and during their consultation visits, so that the mental health providers may have insight into the patient's general quality of life and to track patients' progress in each consultation. A screening questionnaire was included. The major domains of the screening questionnaire are Somatic Concern, Anxiety, Depression, Suicide ideation, Guilt, Hostility, Elevated Mood, Grandiosity, Suspicion, Hallucinations, Unusual Thought Content, Bizarre Behavior, Self-neglect, Disorientation, Conceptual Disorganization, Blunted Affect, Emotional Withdrawal, Tension, Unco-operativeness, Excitement, Distractibility, Motor hyperactivity, Mannerisms and Posturing with responses having 7 options in Likert scales. These structured questionnaires would be administered by primary healthcare providers to help in the identification of the underlying psychopathology of patient's condition. The questionnaires could also be re-administered at various time points, in order for primary care providers to determine whether the patients have been responsive to treatment. In addition, we have integrated the Patient Reported Outcome Measurement Information System (PROMIS) for patients to fill in during each visit to capture General Quality of Life assessment data, in order for primary care providers to recognize the impact of an illness on psychosocial functioning [9]. Furthermore the application aims to provide information that is more specialized for primary care providers and aims to enhance the knowledge base of these non-

psychiatrically trained primary healthcare providers. The information would be updated regularly to include the latest diagnostic criteria for various psychiatric conditions as well as to incorporate treatment plans according to management guidelines based upon the existing NICE (UK) guidelines. Furthermore, psychotropic medication information has also been included in the application for ease of reference, for primary healthcare providers. As psychotropic medications are highly specialized, it is important to provide primary health care providers with accurate and up-to-date information regarding the indications of usage and core side-effect profiles. The mobile application also enables the integration of resources via a resource locator for providers to know about the sub-specialist services located in their areas to allow them to refer patients for other psychological therapies or counseling support, as very often treatment involves both pharmacological and psychological modalities. The mobile application also enables health care providers to send mass e-mails and instant messages to remind patients of appointments or relevant workshops or talks for better condition related knowledge or insight.

b. System design

Content and functionality of the smartphone application was based on the requirements formulated initially. The design of the user interface was designed to be consistent across the various navigation tabs as well as to be easy to navigate with a standard smartphone. Figures 1 to Figures 6 show the individual components of the application.

Figures 1 to 6:

Mental Health

Wellness

[Assessment & First Aid Guide]

for *Health Care Provide*

Dr. Melvyn WB Zhang
Dr. KK Mak

Dx Quiz

Pt's MH

Information

Meds

Help

Assessment Toolkit for Primary Care Providers

Thank you for using our in-house assessment toolkit.

This assessment toolkit has been modified based on then well validated Brief Psychiatric Rating Scale.

This assessment guide serve as a general guide for the initial assessment of the following symptomatology - Depression & Anxiety.

Disclaimer: This assessment toolkit serves only as a general guide.

It is not clinically diagnostic. Individuals are still encouraged to seek help from

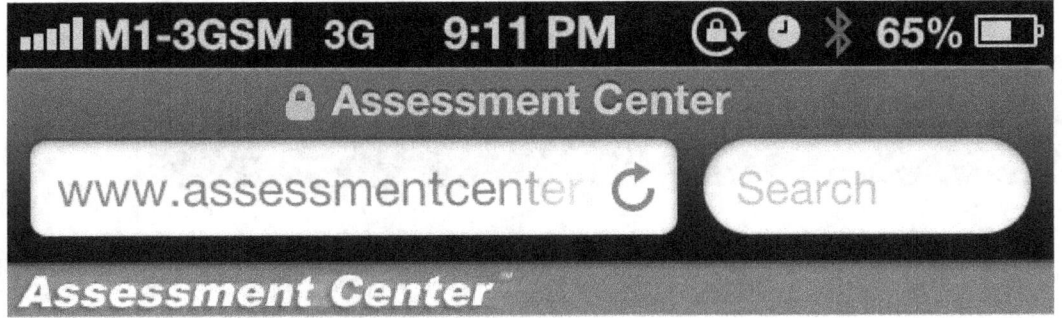

Mental Health Wellness Mobile Application

Individualized Mental Health Wellness

Screening Questionnaire

(please take note of your individualized participant number for record purposes)

If you already have a Login and Password, please enter them in the boxes.

Login   [   ]

Password   [   ]

If you are a first time user, click Start below.

     (Start)

If you have any questions or problems, please contact melvynzhangweibin@gmail.com.

If you would like to bookmark this page, click this **Add to Favorites** link.

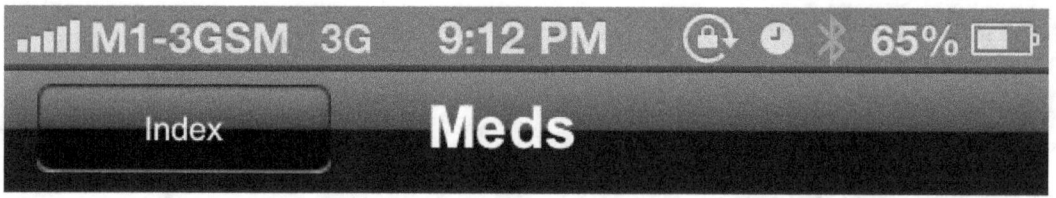
Fluoxetine [Prozac] (Dose range: 20 ñ 60mg/day)

Special features 1) Non linear elimination kinetics 2) safe in overdose.

Other indications: 1) OCD (>60mg/day); 2) Panic disorder; 3) Bulimia Nervosa; 4) PTSD; 5) Premenstrual dysphoric disorder; 6) Premature ejaculation and 7) Childhood & adolescent depression.

Pharmacokinetics: 1) Fluoxetine inhibits the P450 3A3/4, 2C9, 2C19 & 2D6 and it also inhibits its own metabolism. 2) Due to non-linear pharmacokinetics, higher doses can result in disproportionately high plasma levels and of some side-effects (e.g. sedation) rather late in the course of treatment with this drug. Its metabolite norfluoxetine is much less

c. System development

System prototyping was adopted as our software engineering model. The prototype version was built using an online smartphone application builder utilizing HTML5 codes to program specific features in the application within the periods of January 2012 to April 2012. In addition, an online database management system was also utilized to capture the required data.

d. System Evaluation

In this phase, the usability and accuracy of the content was evaluated by an expert group, which comprised of a consultant psychiatrist, a psychiatric resident in training and also an epidemiologist. Each aspect of the smartphone application was repeatedly tested several times to ensure that there are no underlying user-related problems. The robustness of the prototype version was also tested on both the IOS and Android platforms. The consultant psychiatrist, the psychiatric specialist registrar and the psychiatric resident helped to ensure that the content is accurate and updated. The perceived ease of usage of the smartphone application was also evaluated.

e. System Application

The authors considered several mechanisms for deployment of the smartphone application. They have identified that a technology review in a local newspaper might be appropriate – given that it has the widest outreach and also in consideration that most Primary Care doctors are in private partnerships. In addition, an online qualitative survey was created to access the applicability of the application, by asking a group of second year psychiatry residents (n=6) in training to evaluate the application. This group of psychiatry residents was chosen because they have at least one year as trainees in psychiatry, and have completed core medical and surgical posting within the past year. Therefore, they may be presumed to provide more insights regarding the usability of the smartphone application as they are not yet too sub-specialized. Multiple choices based questionnaire was provided for, and key results highlighted as follows.

Results

The pilot prototype was recently featured on the technology section of the Hong Kong South China Morning Post in June 2012 [10] and it yielded a total of 136 downloads during the week when it was featured. The qualitative online survey yielded a response rate of 50%. All the participants felt that it was essential for primary healthcare providers to have access to information in the field. Concerning the importance of information available within the smartphone application, mental health information and medications information were ranked higher than tools like assessment toolkits, resource locators and also the text messaging system to remind patients of crucial events and appointments. Suggestions were given to improve the current application, such as to incorporate medical records information from the electronic medical record database from the local governmental agencies, as well as to further simplify the information for common mental health disorders.

Discussion

Major challenges in application development

There are several challenges with regards to the development of such an application.

1. Meeting the needs of Primary Healthcare Providers

The major challenge involved with the current application development was to fulfill the un-met needs of primary healthcare providers, who are usually the first providers from whom at-risk individuals seek help. This challenge is resolved not only by recognizing the limitations of existing applications, but also by taking into consideration the exact requirements of the providers. We have included the diagnostic questionnaire in order to help primary healthcare providers determine the underlying psychopathology and to help them with their assessment. In addition, we have included individualized Patient Reported Outcome Measurement Information System (PROMIS) "General Quality of Life questionnaires, to enable primary care providers to recognize the extent to which an illness has affect psychosocial functioning and to determine to what extent their interventions, whether pharmacological or psychological, have helped during subsequent consultations. As primary healthcare providers are mostly not psychiatrically trained, the provision of up-to-date information about mental health illness to help them with assessment and diagnosis should be crucial, although this is not a unique feature specific to this application, as there are already existing applications that could provide basic information. The advantage this application over current healthcare information applications might be the incorporation of relevant guidelines such as the National Institute of Clinical Excellence (NICE) guidelines, as well as relevant local mental health guidelines, published by local governmental ministries or departments of health. The provision of accurate, updated information has been acknowledged in our qualitative survey to be of importance to primary healthcare providers.

We also recognize that there are existing applications that are capable of providing detailed information about psychotropic medications. However, a unique feature of our current application is that it provides not only detailed information about these medications, but also the approximate costs of medications available locally to the psychiatric practice, which would be useful because medical care in the local context is not covered by insurance, and at times, patients do have difficulties in affording the newer generations of psychotropic medications. In addition to the side effect profiles of the medications, we have incorporated recommended relevant biochemical investigations necessary for primary healthcare providers to consider prior to commencement of the psychotropics. We have also included recommendations pertaining to the frequency of regular biochemical monitoring as well because certain psychotropic medications, such as mood stabilizers like Lithium and Sodium Valporate would require routine serum levels and other biochemical investigations.

Incorporation of a resource locator enables primary healthcare providers to look easily for alternative modalities of treatment for patients, in conjunction with their existing pharmacological management. The notification system would be useful for primary healthcare providers to remind patients about routine follow-up.

2. Selection of an appropriate questionnaire for Primary Care Usage

There were initial challenges concerning the selection of a questionnaire that would be appropriate for primary healthcare providers to use to help in assessment and diagnosis of a constellation of psychiatric conditions. We have included a simplified assessment questionnaire for quick administration in clinic settings. We have included the validated PROMIS system to help acquire patient's general well-being.

3. Accessibility of Mobile application

Another challenge would be related to the accessibility of the application. Most smartphone users are using smartphones running on either the Apple IOS or the Android operating system. Nevertheless, we have tried to create an application that would be capable of running across more different platforms such as Windows mobile and Symbian, to include more potential users. The prototype model is a web-based application that is able to run across several platforms, inclusive of both the Apple IOS and also the Android platform. Rigorous testing has also demonstrated the ability of the application to be run on mobile tablets in addition to smartphones, for a better user experience. As the screen dimensions of tablets are larger, it would be easier for primary care physicians to fill in the various diagnostic tools as well as to look up information within the application.

4. Security and access issues

We recognize that the current application could be accessed and downloaded by all members of the general public. Because there is specialized medical information within the application, the mental health information and medications resources all require password access. Passwords would only be issued strictly to validated primary healthcare providers.

Actual Barriers

Although the four main challenges have been presented with feasible solutions, the main barrier perhaps would be getting primary care providers to engage and utilize the application in their day to day practice. Some primary care providers might be too overwhelmed with their patient load to have the necessary time to administer the questionnaire scales, check out relevant information and counsel patients accordingly. There is a possibility that they would do a quick clinical interview and refer the patient over to other mental health specialist services. In addition, the majority of the primary care providers currently practicing grew up and studied in a decade to which these advances in technologies have not been made available; and hence, there might be resistance towards adopting this application as a clinical tool.

Limitations

MHWell application has inherent limitations. Although MHWell application is one of the latest innovations in the field of mental healththe number of primary health care providers who would use the application is uncertain since there are already numerous existing resources available online, ranging from websites to various health-related smartphone applications. From the pilot study that was done in Hong Kong, the low download rates of the application could be due to a variety of reasons, such as primary care providers' reluctance to try out new technologies, missing out the news feature, and also inappropriate mechanism used to promote a healthcare related application. The pilot test launch stemmed from an application review published in the Hong Kong South China Morning Post. We recognize the limitations of this approach to publicizing a mobile application for pilot testing. In addition, the qualitative analysis of the application was based on the feedback obtained from 3 Psychiatric Residents. It would have been ideal to acquire more feedback from residents, who have just completed their medical training and have started out on psychiatry and had some psychiatric knowledge. However, this is limited by the low intake rates of the local residency program. Also, in order to sustain the viability of the application, renewal and enhancement of the application needs to be done at least every six months. While most of the younger generation of practitioners are familiar with mobile applications, promoting mental health and completion of smartphone based questionnaires during consultations may be difficult for the older generations due to their technological unfamiliarity. In addition, the current resource locator system is primitive in design, as it simply listed a number of services available in a particular area. A potential further development would involve the use of global positioning technology to enable smartphones to redirect primary healthcare providers to services located nearer to their areas of practice.

Conclusions

Healthcare related applications are rapidly developing, but most of them focus on primary healthcare prevention at an individual level. The development of the Mental Health Wellness application demonstrate how applications could potentially help to enhance primary health care intervention and prevention by targeting mainly primary healthcare providers. There are unique functionalities within the application that enables patients to engage in the process of seeking consultation from their primary care providers, thus providing more comprehensive psychiatric assessment. We acknowledge that our current evaluation is limited, due to the small sample size. It is hoped that the development of the MHWell application could continue and be further evaluated amongst the targeted group of primary care providers, and if proven successful, it would serve as the foundation and basis from which more healthcare related applications could be developed for health care providers involved in primary prevention of common chronic illnesses. The key reason to build the current prototype is to demonstrate the feasibility of building an application catered for primary care providers.

References

1. Mobile Future. Social Media, Apps, and Data Growth Headline 2010 Mobile Trends. Retrieved from http://www.mobilefuture.org/content/pages/mobile_year_in_review_2010?/yearendvideo 2010

2. Blaya, JA, Fraser, HS, Holt, B. E-health technologies show promise in developing countries. *Health Aff (Millwood), 2010;29*(2), 244-251.

3. McClernon FJ, Roy Choudhury R. I am your smartphone and I know you are about to smoke: The application of mobile sensing and computing approaches to smoking research and treatment. *Nicotine Tob Res 2013, May 23*[Epub ahead of print].

4. Pop-Eleches, C, Thirumurthy, H, Habyarimana, JP, Zivin, JG, Goldstein, MP, de Walque, D, et al. Mobile phone technologies improve adherence to antiretroviral treatment in a resource-limited setting: a randomized controlled trial of text message reminders. *AIDS, 2011;25*(6), 825-834.

5. Free C, Phillips G, Felix L, Galli L, Patel V, Edwards P. The effectiveness of M-health technologies for improving health and health services: a systematic review protocol. *BMC Res Notes, 2010;3*, 250.

6. Fioredelli M, Diviani N, Schuiz PJ. Mapping mHealth Research: A decade of Evolution. J Med Internet Res 2013; 15(5):e9.

7. Morris ME, Aguilera A. Mobile, social, and wearable computing and evolution of psychological practice. *Prof Psychol Res Pr, 2012 Dec; 43*(6):622-626.

8. Reeve, BB, Hays, RD, Bjorner, JB, Cook, KF, Crane, PK, Teresi, JA, et al. Psychometric evaluation and calibration of health-related quality of life item banks: plans for the Patient-Reported Outcomes Measurement Information System (PROMIS). *Med Care, 2007;45*(5 Suppl 1), S22-31.

9. Cella, D, Gershon, R, Lai, JS, Choi, S. The future of outcomes measurement: item banking, tailored short-forms, and computerized adaptive assessment. *Qual Life Res, 2007;16 Suppl 1*, 133-141.

10. South China Morning Post. A Hand Tool for assessing your state of mind. Retrieved from http://www.scmp.com/article/1003673/handy-tool-assessing-your-state-mind, 2012 [assessed 2013-02-22]

Chapter 6: Application of Internet and Smartphone technologies for research

Formulation of a Smartphone Application for Crisis Research:

Introduction

Over the past few years, along with the developments in Internet based technologies, there has been much research into using Internet-based platforms largely for the dissemination as well as the collation of medical related information. This has led to the growth of a sub-specialized field, termed as "Infodemiology", which is defined as the science of dissemination of information on an electronic medium, particularly the Internet or in a population with the resultant goal of educating and informing the general public, as well as to help governmental organizations with policy making and planning [1]. In contrast, "Infoveillance" refers to the collation of infodemiology measures for the purpose of surveillance as well as analysis of trends. This sub-specialized field looks at information freely available on the Internet, which could potentially include not only search engines data, but also other forms of data, such as various postings on websites, blogs and discussion boards, as well as data posted on social network mediums like Twitter, Facebook and more [1].

Gunther (2011)[1] has previously outlined the applications of infodemiology in medical research. Data derived from analysis of various search queries on various Internet search engine have been shown to be capable of predicting disease outbreaks [2]. Social network medium like Twitter has provided data that is capable of enabling government and health-related agencies to have faster grasp of the impact of a particular pandemic [3]. It has also helped to identify and monitor the quality of various public health-related information and publications on the internet, thereby enabling earlier identification of significant disparities and errors in health information provided online [1,2]. In addition, it has demonstrated potential as a clinical research tool. It has enabled the development of more patient-centered research instruments that help in monitoring of drug side effects [5,6,7]. It is thus beyond doubts that Infodemiology has added a novel set of tools for clinicians as well as researchers [3]. However despite the advantages that this new form of technology has offered clinicians, there are several limitations. These limitations include the fact that analysis of qualitative information could be tedious, and that there is an inherent sampling bias as the sampling population tends to include those who are younger, more highly educated, with higher average incomes and are more likely to be residing in urban areas [8].

Apart from the developments in Web-based and Internet technologies, there have also been further advances in mobile phone technologies, especially so with the increasing popularity of smartphones. Smartphones are equipped with immense computing abilities and allow individual to utilize Internet always on the go. There has been a massive surge in the number of smartphone applications that are made available for downloading. Statistics have shown an increase from 300 million applications being downloaded in 2009 to over 5 billion in 2010 [9]. In particular, there are currently more than 7000 healthcare related applications made available as of 2011 [10].

Previous research have demonstrated Web-based and Internet technologies in fulfilling their roles in disseminating information during a crisis. Keim and Noji

(2011) [11] stated that the usage of social media for information dissemination first started out during the 2010 Haiti earthquake. Several technological modalities were being utilized in the immediate aftermath of the earthquake, largely to raise funds for those in need, as well as to form support groups for those in need of psychological help. There have also been other research, such as that done by Erza (2012) [12] that have looked into the usage of Face-book for clinical assessment to compare the level of Post Traumatic symptoms experienced during the 2011 Fuskshima nuclear disaster. Previous research has focused much on the evaluation of existing web-based and Internet technologies and not on exploring smartphone based technologies for disseminating and collating crucial information during a major crisis. When Asia was afflicted with the recent Southeast Asia Haze Crisis in 2013, the authors have decided to implement a Haze Crisis application to allow dissemination of pertinent information and collate crucial information on physical and psychological well-being from individuals. **Our objectives for the current study were to demonstrate the methodologies behind building a smartphone based application for dissemination and collation of information in a major air pollution crisis.**

Methodology
The Southeast Asia Haze Crisis Smartphone Applications
The web-based smartphone application was developed and made available between 21st to 26th of June 2013. The end date for the application was decided upon as the haze crisis has improved significantly with the air quality returning to the healthy range by the 26th of June 2013.

The web-based smartphone application was developed based on a developmental approach adopting five phases, which are a) formulation of user requirements, b) system design, c) system development, d) system evaluation and e) system application. The details pertaining to each phase of system development are as documented below.

a. Formulation of user requirements
The Haze crisis application was developed in consideration of the current limitations in the existing field, as there are not many crisis related applications, and also no existing application has been made available specifically targeting any air-pollution crisis. The application aims to provide the general public with updated information pertaining to the current air quality level, so as to enable them to plan their activities accordingly. Furthermore, in consideration of the impact that the haze has on outdoor activities and also with the governmental agencies making recommendations on the levels of activities depending on the current pollutant severity index, an event re-scheduling system was integrated into the application itself. Taking into account the impact that the haze can have on an individual's physical well-being, a list of crucial contact numbers for respective hospital emergency services and twenty-four hours clinics were also included. Previous studies have highlighted that for any healthy individuals, the haze may cause medical complications such as conjunctivitis, throat irritation, rhinitis, blocked nasal passages, excessive sputum production and even significant breathlessness, headache and slow cognition. Those with pre-existing respiratory conditions may find their conditions worsening, and those with dermatological conditions like eczema might also find exacerbation of their skin conditions. In particular, those with pre-existing cardiovascular conditions may suffer from an individual risk of myocardial infarction or stroke and those who are pregnant and asthmatic may also suffer from reduced oxygenation of the fetus. Given the

physical implications such a major air pollution crisis has on individuals, it is also crucial for researchers and particularly local governmental organizations to acquire data relating to the physical and psychological well-being of individuals. Hence, within the mobile application, there should be feedback mechanisms such as surveys that are validated to acquire these data.

The questionnaires that were included to evaluate participant's well-being were written in English and consisted of four main sections, which were (a) Demographics, (b) Physical symptoms experienced during haze, (c) Perspectives on usefulness of healthcare equipment and (d) Impact of Event Scale - Revised (IES-R). An online database coded by using an online website (Polldaddy) was created and used to capture the responses. The demographics questionnaire comprised of 7 items in total and was used to acquire the baseline characteristics of participants partaking in the study, such as gender, age, ethnicity, marital status, level of education, occupation and most importantly, presence or absence of chronic medical illnesses. The questionnaire on physical symptoms enquired the presence or absence of the following physical symptoms: mental slowing, headache, dizziness, eye discomfort, nose discomfort, mouth and throat discomfort, breathing difficulty, heart pain or chest pain, nausea and vomiting, gastric or abdominal discomfort, slowness in movement and muscle ache or pain. Participants rated on the range of self-perceived dangerous PSI values, personal possession and perceived usefulness of the N-95 mask on 5-point Likert scales. The Impact of Event Scale- Revised (IES-R) is a 22-item self-administered questionnaire that has been well validated for determining extent of stress reaction after exposure to stressful circumstances, within one week of exposure across different cultural groups (Creamer et al 2003, Weiss 2007). Each item on the scale was administered via a 5-point frequency scale (0 - Not at all, 1 - A little bit, 2 - Moderately, 3 - Quite a bit, 4 - Extremely) and higher scores indicate higher level of psychological stress. The IES-R provides three sub-scores, the mean avoidance, intrusion and hyperarousal scores, a total mean IES-R score (divided by the total number of items) and a total IES-R score (without division by the total number of items). A total IES-R score equal to or more than 33 signifies the likely presence of post-traumatic stress disorder.

b. System design

Content and functionality of the smartphone application was developed based on the requirements formulated initially. Due to the rapid onset of the air pollution crisis, it was not feasible to engage a software company to build a native IOS or Android application. By making use of an online web-based application builder (www.conduitmobile.com), the authors were able to create the application within 12 hours after the onset of the major air pollution crisis. The design of the user interface was such that it was consistent across the various navigation tabs. The design was kept simple in order to make sure that the application was easy to navigate with a standard smartphone.

Figures 1-5 show the individual screen-shot of the individual components of the application.

c. System development

The prototype version was built using an online smartphone application builder (www.conduitmobile.com), using HTML 5 codes to program specific features within the application. The updated air quality index was cached directly from the local

National Environmental Agency Website and integrated within the application. The event management system for users to reschedule events have been previously developed by the author (MWBZ) previously, and would be capable of sending off text message triggers when users create and modify any of their planned events online. The list of emergency contact numbers of hospitals as well as the contact numbers of twenty-four hours clinics were obtained from the Ministry of Health Singapore Web-site and were hard coded within the application.

In addition, an online database management system (www.polldaddy.com) was used to capture the required data. The online management system that was chosen for usage had the capability of exporting collated data in either the ..csv or .xls file formats.

d. System evaluation

In this phase, usability and accuracy of the content were evaluated by an expert group. Each aspect of the smartphone application was repeatedly tested several times to ensure that there are no underlying user-related problems. The robustness of the application was also tested on both the IOS and Android platforms. The perceived ease of usage of the smartphone application were also evaluated by the expert group.

e. System application

The smartphone application was deployed during the period (as previously mentioned) via a self-sponsored Facebook post, featuring both a direct link to the smartphone web based application and to the questionnaire. Apart from the Facebook sponsored post, emails containing the web-links to the questionnaires were also disseminated to selected group of respondents. There is an underlying rationale for the authors to be making use of Facebook to disseminate the application. Given that the application developed was a mobile application, it cannot be searched via the application stores. Hence, the authors have decided to make use of the existing social media technologies to spread the information about the availability of the application.

www.nea.gov.sg/psi/   ↻   Search

National Environment Agency
Safeguard · Nurture · Cherish

🦁 **Singapore Government**
Integrity · Service · Excellence

About NEA   Career   Newsroom   Feedback   Sitemap   ⓔ **FAQS**   Contact NEA

| Public Health | Anti-Pollution & Radiation Protection | Energy & Waste | Events & Programmes | Grants & Awards | Training & Knowledge Hub | Services & Forms |

Home > Anti-Pollution & Radiation Protection > Air Pollution > PSI > PSI and PM2.5 Readings   🔲 🔲 🔲 ✉ 🖨 A⁺ A⁻

y and Targets
on Regulations

PSI and PM2.5 Readings

d PM2.5 Readings
t Concentrations
xlity of PSI Readings
Suppliers of Air Cleaning
s for Buildings

d Testing Bodies For
mission Tests

Air Quality

the Haze

ful Links

on

n

lution

tection

ng Planning

Qs under this topic.

3-hour PSI Readings from 12AM to 12AM on 21 Jun 2013

Time	12AM	1AM	2AM	3AM	4AM	5AM	6AM	7AM	8AM	9AM	10AM	11AM
3-hr PSI	210	173	143	119	104	96	94	111	158	256	367	400
Time	12PM	1PM	2PM	3PM	4PM	5PM	6PM	7PM	8PM	9PM	10PM	11PM
3-hr PSI	401	360	245	168	-	-	-	-	-	-	-	-

Get more haze info and maps here.

Hourly updates of 3-hr PSI readings are provided from 12midnight to 12midnight. The 3hr PSI readings are calculated based on PM10 concentrations only.

24-Hr PSI Reading on 21 Jun 2013

8 am Reading		12 pm Reading	
Region	PSI	24-hour PM2.5 Concentration (µg/m³)	Health Advisory
North	197	278	The following groups should remain indoors and keep activity levels low: •People with heart or lung disease •Children and older adults. Everyone else should avoid all physical activity outdoors.
South	196	266	The following groups should

◀   ▶   ⬆   📖   🗔 2

www.pointintime.com.sg ↻   Search

Share events easily and effectively!

point in time
post and share events

Good day, Guest! (Sign in)

Events   Suppliers   Create an Account   FAQ

RECENTLY POSTED EVENTS

Post an Event

What event is coming up soon?

It's a... [Anniversaries ▼] (Next)

Got an event code via SMS/Email?

Enter the code below to see the details.

Code [_____] (View)

ADVERTISEMENTS

Recent Event Comments

SUPPLIER DIRECTORY

Bakery	Funeral Services	Photographer
Fashion	Gifts	Restaurant
Florist	Hotel	Wines
Food Catering	Others	

Featured Suppliers

Check out the directory →

Do you have a product or service for event hosts?
Join our directory!

Home   About   Advertise   Rates   Events   Supplier Directory   FAQ   Contact
Us

Copyright 2009 Pointintime Private Limited

Privacy Policy   Terms of Service

     2

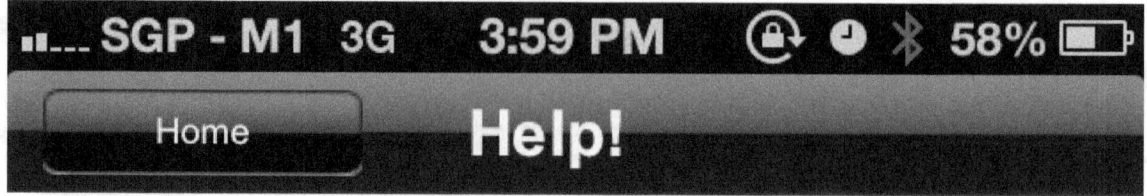

Seeking help:

HOSPITAL 24 HOUR EMERGENCY

Singapore General Hospital

6321-4311

(+65) 6326 5656

(24 hour hotline for international patients)

Tan Tock Seng Hospital

6357-8866

6779-2777 (24 hour - day & night)

Alexandra Hospital

6379-3150, 6379-3840(A&E enquiries)

.ıl.... SGP - M1 3G 4:01 PM 57%

masteringpsychiatry.po Search

Haze and Psychological Wellbeing

Thank you for participating in this voluntary survey.
We wish to understand the impact of the current Haze on your quality of life.
This questionnaire is adapted from:
The Hartford Institute for Geriatric

References

[1] Eysenbach G. Infodemiology and infoveillance: framework for an emerging set of public health informatics methods to analyze research, communication, and publication behavior on the internet. J Med Internet Res 2009; 11(1):e11.

[2] Ginsberg J, Mohebbi M, Patel R, Brammer L, Smolinski M, Brilliant I. Detecting influenza epidemics using search engine query data. Nature 2009;457(7232):1012-1014.

[3] Chew C, Eysenbach G. Pandemics in the age of Twitter: content analysis of tweets during the 2009 H1N1 outbreak. PloS ONE 2010;29(5);e14118.

[4] Eysenbach H. Infodemiology: the epidemiology of (mis)information. Am J Med 2002;113(9):763-5

[5] Wicks P, Massagli M, Kukkarni A, Dastani H. Use of an online community to develop patient-reported outcome instruments: the Multiple Sclerosis Treatment Adherence Questionnaire (MS-TAQ). J Med Internet Res 2011;13(1):e12.

[6] Wicks P, Massagli M, Frost J, Brownstein C, Okun S, Vaughan T. Sharing health data for better outcomes on PatientsLikeMe. J Med Internet Res 2010; 12(2):e19.

[7] Frost J, Okun S, Vaughan T. Heywood K, Wicks P. Patient reported outcomes as a source of evidence in off-label prescribing: analysis of data from PatientLikeMe. J Med Internet Res 2011;13(1):e6.

[8] Statistics Canada. Canadian Internet Use Survey, 2009. www.webcitation.org/5w9gvNaWt.

[9] Mobile Future. Social Media, Apps, and Data Growth Headline 2010 Mobile Trends. Retrieved from http://www.mobilefuture.org/content/pages/mobile_year_in_review_2010?/yearendvideo 2010.

[10] Free C, Philips G, Felix L, Galli L, Patel V, Edwards P. The effectiveness of M-health technologies for improving health and health service: a systematic review protocol. BMC Res Notes, 2010;3:250.

Chapter 7: Future Directions for Psychiatry to embrace technologies

Framework that Psychiatry could adopt for Education

Despite the fact that literature search reveals no recent review of applications specific to Psychiatry for education, there are actually quite a lot of applications for academic purposes on the commercial application stores, such as Apple ITunes and Android Play. In order to systematically analyze the current breath of peer-reviewed literature for education in psychiatry, and to summarize the key findings of current research, and to identify current knowledge gaps, it might be worthwhile to consider a Scoping Review. Scoping reviews have been done by Katherine (2013) recently to assess the usage of technology in delivering mental health service for children and youths in Canada. Descriptive numeric summary and thematic analysis were conducted for the literature reviewed. In summary, the authors in their review concluded that the usage of technology did indeed play a major role in the service and supports to children and youth, and also in terms of prevention, assessment, diagnosis and treatment. Similar scoping review should be adopted to assess systematically the application of technology for Psychiatry as a specialty, as it would be key to understand the current breath of research; and to understand the impact of current research and also to systematically identify the key gaps in education. The scoping review could also consider commercial applications that are currently available on the respective application stores and highlight the common themes these applications cover.

Faye H (2013) has proposed two other methodologies to overcome the issues that there is a lack of applications that has been reviewed by appropriate authorities to demonstrate that they are evidence based. One methodology involves having University or Healthcare organizations creating their own in-house applications. It has been believed that the development of 'in house' applications has its inherent advantages, as the application could help to address current shortfalls in clinical education, or current deficiencies in competencies of either medical students or residents. The other methodology would be for either University of Healthcare organizations to have their own compilations of their own peer-reviewed applications that are deemed suitable for usage by either medical students or residents. This helps to ensure that University and Healthcare organizations provide only high quality, peer reviewed applications that could be recommended to medical students or residents.

Summary of Framework that Psychiatry could adopt for Educational Needs
    a. Scoping Review to identify the current gaps in education for Psychiatry; and to identify current commercially available Psychiatry Educational Applications
    b. Enabling clinicians and educators to develop their own in-house educational applications
    c. Enabling clinicians and educators to identify suitable educational applications through a systematic peer review process

---

Developing own in-house educational applications
With regards to the application of the first methodology for Psychiatry education, clinicians from Universities could play a larger role by making use of simple methodologies to develop simple web-based applications. This has been previously

described and illustrated by Yousif (2014), in their article entitled "Designing web-apps for smartphones can be easy as making slideshow presentations." In their article, they described how a simple web-based mobile application could be made using just an Internet browser and also a text editor. Using the application online known as jQuery Mobile Website (based on Codiqa interface), the clinician will initially begin at the main frame and then be directed towards the other frames. Integration of connections between the frames could be done using a simple drag and drop methodology. The online interface will automatically generate codes based on the content that the clinician has included. The clinician could collate the codes, paste them into a text-editor file and then integrate 2 other codes at the top and the bottom of the page; and a simple smartphone based web application would have been created. Hence, using the methodology as described, clinicians will be able to create their own smartphone applications for dissemination of information they feel is pertinent for the educational needs of medical students and residents.

Identifying suitable educational applications through a systematic peer review process

The other methodology that has been proposed will be for Universities and healthcare organizations to have a collection of peer reviewed applications that are deemed to be suitable for usage by educators, clinicians as well as by their patients. Previous research has proposed utilization of a systematic self-certification model for app review.

Thomas LL (2013), in his recent editorial reply has proposed the application and utilization of a systematic self-certification model for peer review of applications. It is considered to be quicker, and it offers a systematic solution for clinicians and healthcare professionals to have a systematic way to judge and determine whether an application is safe and is applicable to their practice. The self-certification model for peer review of applications builds on what has been previously developed by the Health on the Net foundation (HON), for assessment of the reliability and credibility of the information present on medical and health websites. The following table illustrates aspects of the proposed model:

Criteria	Description of Criteria
Nature of Information present in application	All relevant medical information that is included within the application must be attributed to an author. The level of the training of the author in his or her specialty must be cleared presented within the application
Purpose of the application	There should be a statement documenting the main purpose of the application as well as the intended target audience.
Confidentiality	It is mandatory that the application document a privacy policy as to how confidential, private or semi-private information such as email address and the contents of the emails received are treated. The application has to inform users whether their data will be recorded in any databases.
Information	It is essential that all medical information embedded within the application have a specific date of creation, a last modification date and there must also be appropriate references.
Justification of claims	If there are information about the benefits or performance of any treatment (whether medical or surgical), it has to be supported by concrete scientific evidence.
Contact details	There must be inclusion within the application a specific way to

	contact the developer.
Disclosures	Applications should disclose their sources of funding, for example, governmental agencies, private companies or donations. Developers need to also state their conflicts of interest.
Advertising policy	Applications that display paying banners need to have an advertising policy. The policy need to explain how the application distinguishes between editorial and advertising content and which advertisements are accepted.

Adapted set of criteria based on the HONcode

The above self-certification model might be of use to Psychiatry. Psychiatrists could make use of the above model for peer review of relevant applications in their specialty. They will be able to identify specific applications (whether education or clinical) that could potentially be of value in their specialty. With regards to education, perhaps an additional criteria to be considered will include the extent of applicability of the application to their specific organization, as very often, there are epidemiological data and guidelines that might not be applicable to a particular country.

This is also in view of the fact that there has been, over the recent years, more psychiatry related applications in the respective application stores. It is thus key for Psychiatrists to have a model to evaluate applications and identify those that are currently safe and beneficial for their practice, prior to them developing new applications, to fulfill current needs and knowledge gaps in their specialty. By doing so and with more publications in leading informatics journal about peer reviewed psychiatry related applications, psychiatry as a specialty will further advance, as the prior review conducted demonstrated that most specialties are limited to identification of specific applications and has not embarked on using models for peer review and evaluation.

Framework that Psychiatry could adopt for Patient's Care

There are several proposed methodologies that may overcome the previously identified limitations. A current limitation is the general lack of evidence regarding the clinical efficacy of psychiatric smartphone applications. These applications have great potential in terms of improving patient care and clinical practice, and no potential disadvantages associated with their use have been identified. Previous research has suggested that peer review of medical applications, in addition to provisional testing by relevant patient groups, could be one method to address these concerns (Buijink, 2013).

a. Peer Review of Psychiatry Healthcare Applications Using Randomised Controlled Trials
With the advancements in smartphone technology and the limitations of the existing literature, we recommend that psychiatrists provide more frequent literature reviews of applications pertinent to their practices. Such reviews would help to identify a collection of psychiatry applications that have been useful in patient care, optimally in the areas highlighted above. Psychiatrists should also conduct randomised controlled trials using these applications (similar to the manner in which previous psychiatric applications have been reviewed) and publish the results demonstrating the efficacy of

these online applications. As randomized controlled trials usually take time to complete, a proof of concept or a feasibility study might be sufficient for an application that has a basis on existing conventional methodologies.

b. Identification of a Collection of Safe Healthcare Applications Using a Systematic Self-certification Model

Alternatively, psychiatrists or healthcare organisations could compile a list of peer-reviewed applications that are deemed appropriate for use by healthcare professionals and patients. Previous studies have proposed the utilisation of a systematic self-certification model for application review. Thomas LL (2013), in his recent editorial reply, proposed the application and utilisation of a systematic self-certification model for the peer review of applications. This method would rapidly provide a systematic solution for clinicians and healthcare professionals who require a systematic method to determine whether an application is safe and applicable to their practice. The self-certification model for the peer review of applications builds on previous developments by the Health on the Net foundation (HON) for the assessment of the reliability and credibility of the information presented on medical and health websites. The following table illustrates certain aspects of the proposed model:

Criteria	Description of Criteria
Nature of Information Present in the Application	All relevant medical information included in the application must be attributed to an author. The level of training of the author in his or her specialty must be clearly presented within the application.
Purpose of the Application	There should be a statement documenting the main purpose of the application and the intended target audience.
Confidentiality	It is mandatory that the application document a privacy policy regarding how confidential, private or semi-private information such as email addresses and the contents of emails received are treated. The application must inform users whether their data will be recorded in any databases.
Information	It is essential that all medical information embedded within the application have a specific date of creation, the most recent modification date and appropriate references.
Justification of Claims	Any information regarding the benefits or performance of any treatment (whether medical or surgical) must be supported by concrete scientific evidence.
Contact Details	There must be a specific method for contacting the developer, and this information must be included in the application.
Disclosures	Applications should disclose their sources of funding, for example, governmental agencies, private companies or donations. Developers must also disclose any conflicts of interest.
Advertising Policy	Applications that display paid banners must have an advertising policy that explains how the application distinguishes between editorial and advertising content and which advertisements are accepted.

Adapted set of criteria based on the HON code.

c. Peer Review and Identification of Safe and Useful Applications by Government Organisations

In addition, specific organisations within the United Kingdom should spearhead the review of psychiatry applications. The NHS Choices health applications library (http://apps.nhs.uk), which was initiated in December 2012, aims to accomplish this goal by identifying a collection of applications and having them reviewed by health professionals, application developers and the general public. Sustaining this rigor in the evaluation of applications will result in a list of safe and trusted applications. Currently, 17 mental health applications have been identified and listed as safe in the application stores. This collection may be useful for practicing psychiatrists to recommend to their patients.

Chapter 8: Enabling Psychiatrists to be Smartphone App Developers: A Concise guide

Insights into app developmental methodologies
Smartphone application development

There are currently 2 methodologies that could be used to create HTML5 mobile web-based applications, without the need of any technical programming knowledge. The first methodology is via using the portal (www.wordpress.com)[7], which has been commonly known to the general public as a blogging site. When it is being applied to medicine, all clinicians would need to do is to create an account, register for a domain name, and modify the content using the graphic user-interface that it offers [Figure 1].
--------------------------- Insert Figure 1---

Figure 1: Methodology of creation a web based HTML5 smartphone application using Wordpress

Text-based content integration is possible by dragging and dropping the appropriate content into the posts or pages. Multimedia features like videos could be uploaded to the library and they would be automatically integrated into the page. Forms could be generated and filled forms could also be directed to clinician's email. When the portal is launched on a normal computer, it would be a full-fledged website, but on a smartphone, it will be automatically displayed as an application.

The other methodology of developing a web-based mobile application is by using online mobile web-based application builders like Conduit mobile [8]. Its graphic user interface would help in the immediate integration of text-based content, videos and even RSS feeds (for clinicians to retrieve information from a dedicated server).

-------------------------- Insert Figure 2 --

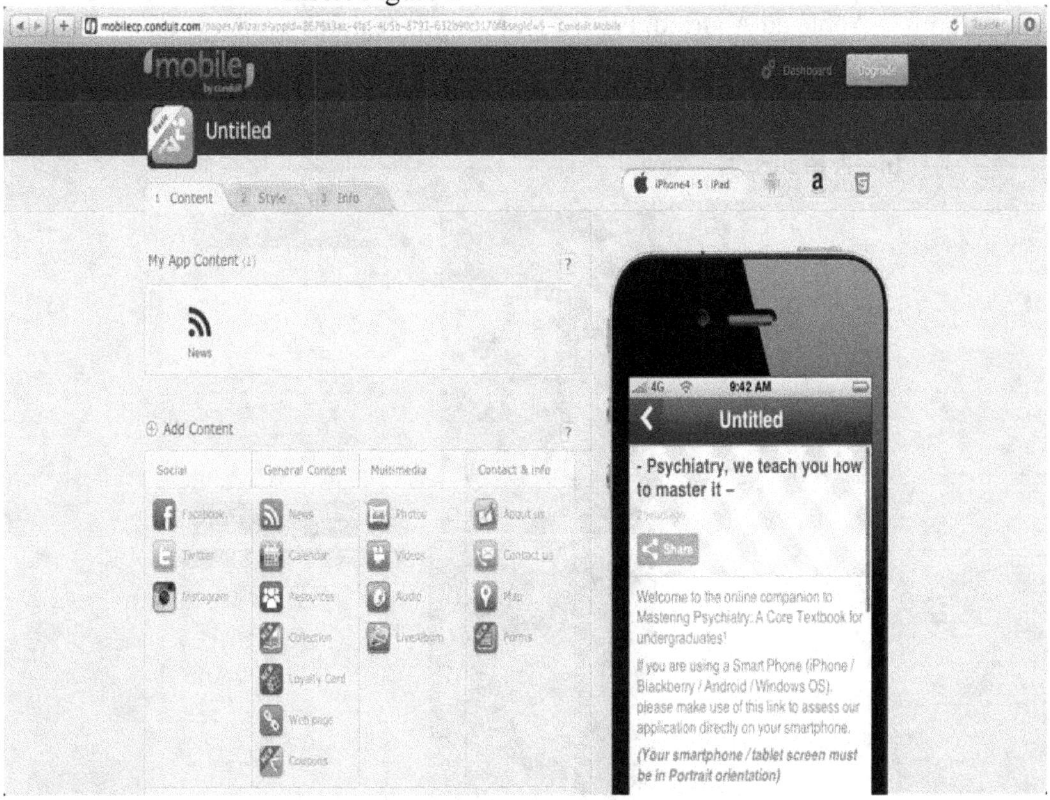

Figure 2: Development of Web-based smartphone application using online application builder.

The two methodologies described above enabled clinicians to devise cost effective applications. Using Wordpress.com to build an online portal and application is entirely free, and would cost USD18 to purchase a specific domain. Using the online application builder to build the smartphone application also in itself does not involve any cost. However, there is a restriction that a white advertisement label will appear on the application. Hence, the authors will describe the development of two applications using the cost-effective methodologies as described and evaluate the receptiveness of users towards these cost effective applications.

The Mastering Psychiatry Online Portal and Smartphone Application (Web-application)

The web-based portal and web based smartphone application was developed between February 2012 to June 2012 using the above methodology. The developmental approach involved five core phases, which included that of: a) formulation of student's requirements, b) system design, c) system development, d) system evaluation and e) system application. With regards to user's requirements, the authors have identified the following to be essential: integration of core textbook information, psychiatry clinical skills interviewing videos, multi-choice questionnaire to assess domain knowledge as well as a patent-pending text messaging system that enables students to schedule events like tutorials and be reminded via text messaging. The determination of the relevant content to be included was based on the experiences of the authors in education; and also based on careful exploration of existing textbook based applications on the app store. System evaluation was conducted by the authors across 2 main platforms – Apple ad Android. Figure 3 shows the features available on the online portal as well as within the application.

---------------------------- Insert Figure 3---

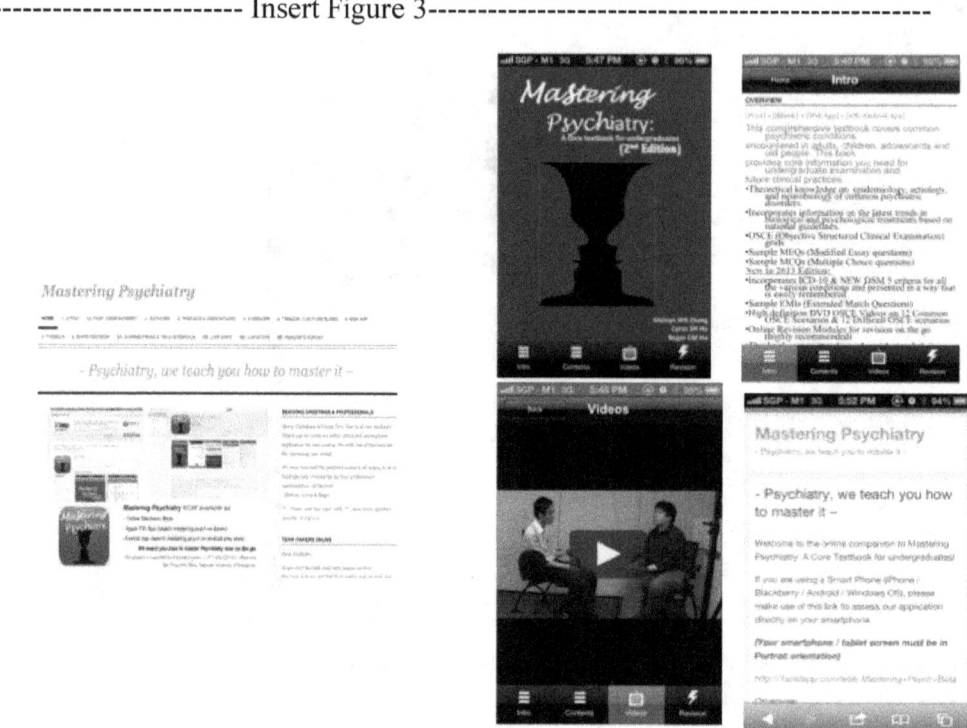

Figure 3: Core features integrated within the Mastering Psychiatry Web-based Smartphone Application

The Déjà vu CASC Application (Web-based smartphone application)

The Déjà vu CASC application was developed between January 2014 and April 2014. The developmental approach involved 4 developmental phases, which included that of: a) Understanding trainee's requirements, b) System development, c) System evaluation and d) System deployment. In terms of user's requirements, the authors postulate that the core requirements of a CASC application for trainees would need to include each one of the following:

1. Inclusion of Mock examination stations with timers (either 7 minutes or 10 minutes timers)
2. Timers to include additional 60 seconds preparation time, to allow trainees to practice recalling vital information and write them down prior to the commencement of any station
3. Inclusion of mock stations (30 in total) that are adaptations of old College stations with variants in the constructs, or are new stations crafted based on information from the Royal College of Psychiatrists Mental Health Leaflets

4. Ability of application to enable trainees to link up with others for video-conferencing
5. Ability of application to enable trainees to practice stations in a timed mock examination way with fellow trainees from their trust
6. Inclusion of instructional videos that demonstrate to trainees core approaches for a specific variety of stations

The postulation of what to be included in the application was based on the recent experiences of the authors in the recent diet of the postgraduate examination.

In consultation with the author RCMH, the following core interviewing approaches videos have also been identified to be essential:
1. How to break bad news
2. How to deal with angry patients
3. How to perform a mental state examination
4. How to handle an explanation station
5. How to deal with disinhibited patients
6. How to deal with patients who are refusing to engage with Psychiatry
7. How to perform a risk assessment
8. How to deal with patients who are not forthcoming
9. How to handle patients with learning disabilities
10. How to perform cognitive assessment

Figures 4 to Figures 7 show the features implemented within the application.
----------------------------- Insert Figures 4 to 7---

Figure 4: Overview of the application

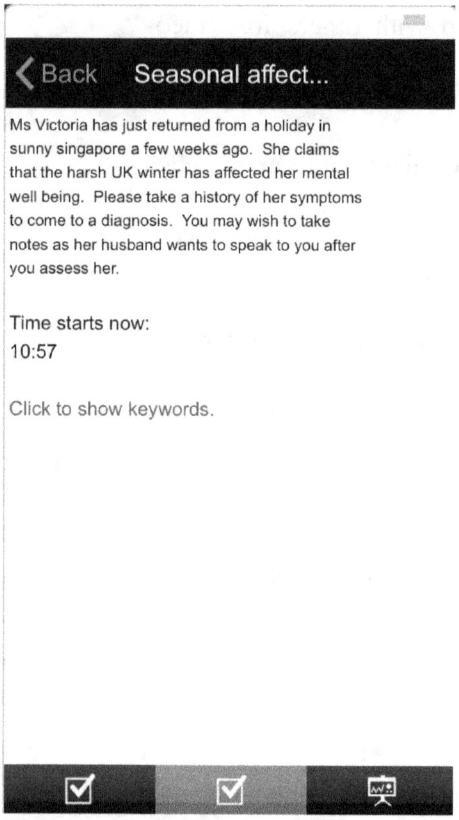

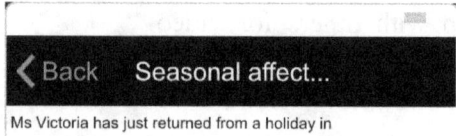

Back
Seasonal affect...
Ms Victoria has just returned from a holiday in sunny singapore a few weeks ago. She claims that the harsh UK winter has affected her mental well being. Please take a history of her symptoms to come to a diagnosis. You may wish to take notes as her husband wants to speak to you after you assess her.
Time starts now:
10:57
Click to show keywords.

Figure 5: Overview of a Mock Station

Back
Seasonal affect...
Ms Victoria has just returned from a holiday in sunny singapore a few weeks ago. She claims that the harsh UK winter has affected her mental well being. Please take a history of her symptoms to come to a diagnosis. You may wish to take notes as her husband wants to speak to you after you assess her.
Time starts now:
10:54
Keywords: -assess circumstances -core symptoms of depression -biological symptoms (increased appetite and sleep!) -cognitive symptoms -emotional symptoms -risk assessment (suicidal intent/plans) -comorbids (psychosis, drugs/alcohol) -significant personal hx (past psych hx, family hx)
Click to show keywords.

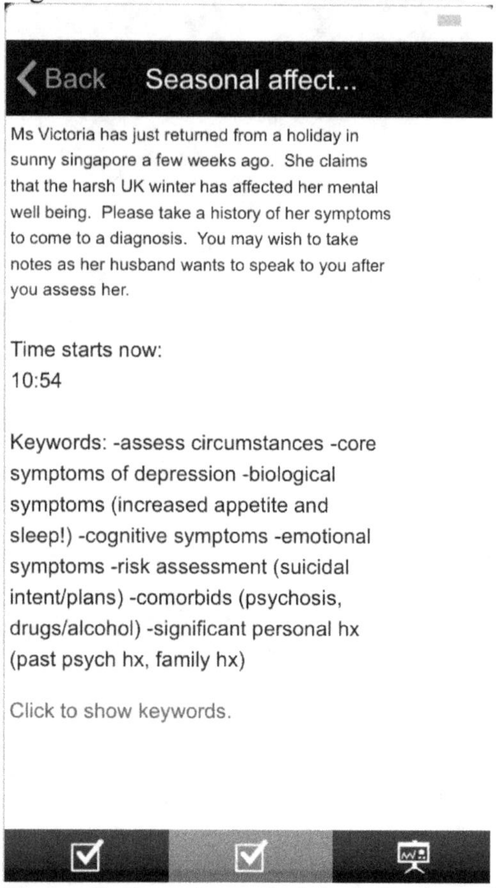

Figure 6: Overview of a Mock Station with keywords displayed

Figure 7: Overview of Video integrated within application

References:
1. Wordpress http://www.wordpress.com
2. Conduit Mobile http://www.conduitmobile.com

www.ingramcontent.com/pod-product-compliance
Lightning Source LLC
Chambersburg PA
CBHW080825170526
45158CB00009B/2530